D1823739

AUSTRALIA THE BEAUTIFUL
WILDERNESS

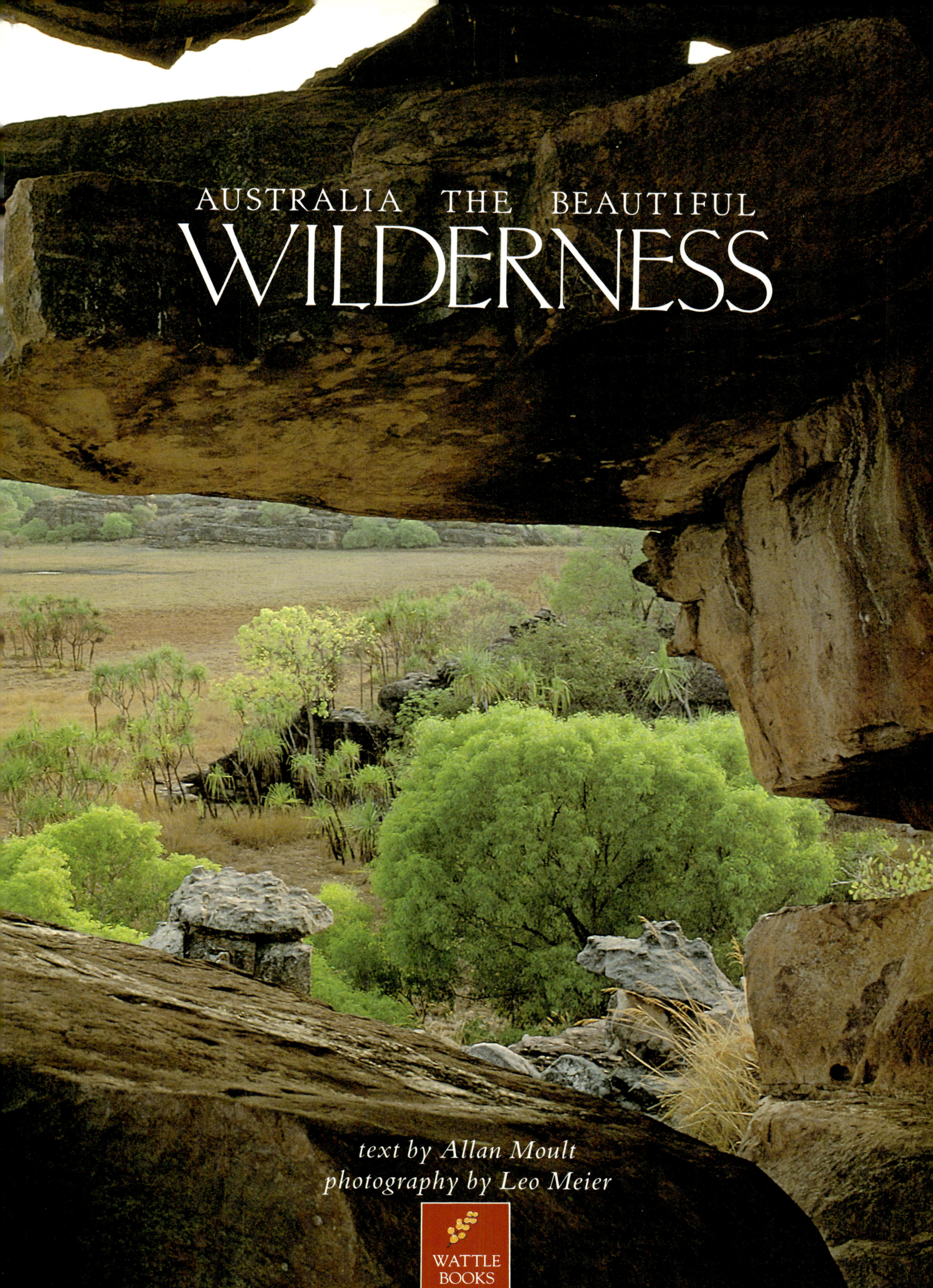

AUSTRALIA THE BEAUTIFUL
WILDERNESS

text by Allan Moult
photography by Leo Meier

WATTLE
BOOKS

First published 1983 by
Kevin Weldon and Associates Pty Ltd,
43 Victoria Street, McMahons Point, NSW 2060, Australia
© Copyright Kevin Weldon 1983

All rights reserved. No part of this publication may be reproduced,
stored in a retrieval system or transmitted in any form or by any means
electronic, mechanical, photocopying, recording or otherwise
without prior written permission of the publisher.

Project Co-ordinators: Kerrie Andrews and Elaine Russell
Design and Art Direction: Tony Gordon
Finished Art: Paul Geros
Typesetting: Walter Deblaere and Associates
Photolettering: The Dova Type Shop

KAKADU NATIONAL PARK,
Northern Territory

HINCHINBROOK ISLAND
NATIONAL PARK, Queensland

A U S T R A L I A

THE SIMPSON DESERT,
South Australia

WASHPOOL,
New South Wales

FITZGERALD RIVER
NATIONAL PARK, Western Australia

BOGONG
NATIONAL PARK, Victoria

SOUTHWEST NATIONAL PARK, Tasmania

National Library of Australia Cataloguing-in-Publication Data
Moult, Allan
    Australia the beautiful, wilderness.
    Includes index.
    ISBN 0 949708 02 X.
    I. Wilderness areas – Australia.
    I. Meier, Leo. II. Title.
    (Series: Wattle books.)
333.78'2'0994

Photography on Tasmania by Allan Moult

*Produced by Mandarin Publishers Ltd, Hong Kong
for Kevin Weldon and Associates Pty Ltd, Australia*

CAPTIONS
*p 2-3, Hinchinbrook Island National Park;
p 4-5, Bogong National Park; p 6-7, Simpson Desert;
p 8-9, Kakadu National Park; p 14, Washpool.*

ACKNOWLEDGEMENTS

*This book could not have been completed without the enthusiastic support of the rangers and staff of Australia's National
Parks and Wildlife Services. Special thanks must also go to Keith Bradby and Heather Pearce, for showing us the secret places
of the Fitzgerald River National Park, Greg Miles and Ian Morris for sharing their knowledge of Kakadu National Park,
Milo Dunphy for sharing his passion for our wilderness areas, Ian May, Jarovslav Pecanek and Paul Chard for guiding us
safely across the Simpson Desert and Bill Albion for stamina and good humour during our 20-day hike to the South West
Cape of Tasmania. And of course there are those friends who shared various journeys and brought laughter and infectious
enthusiasm to our project. Thank you Kathie Atkinson, Andrew White, Ken Mextel, Stephanie and Ian Bowman.
The publisher wishes also to thank Bell & Howell, Australian distributors of Nikon cameras, for their kind co-operation
and assistance.*

Contents

Foreword

TWO HUNDRED YEARS AGO Australia, the world's smallest continent, was all wilderness. Its people lived in wilderness. They were part of it. Australia was the same as it had been ten thousand years ago when, before the advent of agriculture and industry, the whole world was wilderness and everyone lived in wilderness.

Now only a few small patches of this vast Australia remain wild and unchanged. Two thirds or more of the country's original forests have been axed and nearly all of the remaining forests are earmarked for chipboard, paper bags or newspaper stock. All but a couple of Australia's great rivers are dammed, silted or polluted. Cities, farms and factories cover the most fertile areas. The air most people breathe carries car fumes, chemicals and the dust of overgrazed, arid inland plains.

Especially in the south, Australia's original wilderness-bred people have been mostly done to death or deprived their land and lifestyle, and abandoned. Their natural wisdom has been ignored by the march of modern materialism.

Millions, billions, trillions of imported cats, camels, cockroaches, cattle, carp, cacti, rabbits, starlings, sheep, thistles, blackberries, people, pines, foxes and toads are at work destroying the natural fabric of Australia.

The invading European pioneers may be allowed an excuse for having begun all this – they knew no better. Many earnestly thought that they were the chosen people of a God who had given them 'dominion over every moving thing that liveth upon the earth' *(Genesis 1.28)*. Others were simple desperadoes deprived of their origins, family, dignity and security. For these wretched folk, mainly convicts, soldiers or dispossessed peasants, this alien land offered little more than a struggle for survival.

Many, in fact most, of the new arrivals had no understanding of the wilderness continent or of the wilderness wisdom of its original people. When people do not understand, they do not respect and often they fear. From fear comes destruction. The Aborigines with their wilderness homelands were slaughtered. Even so, a few of the new Australians trod softly and even wept for the way their new country was treated. But they were few.

No doubt most of us are better off for the hardship and work of the pioneers. From their sweat has come our prosperity. We are healthier, wealthier and more secure in life than they would have dreamed possible. We have education, mobility, opportunity and time to think. We also know more about our planet and our country. We have an overview of Australia which they never had.

In fact, if the drastic changes to the natural face of Australia were all in the past, we could perhaps commemorate them and be glad that we were now doing better. But the depravations have, if anything, increased. We are doing worse. And it is all the worse because we now know what we are doing in a way that our forebears did not. Unlike them, we know how little wilderness is left. We therefore have an unprecedented responsibility to care for that wilderness.

Modern Australians have no excuse for letting the plunder go on. The responsibility we have is all the greater because it is tied into the global community's greatest problems – the starvation facing millions of people, the imminent threat of a nuclear holocaust and the cruelty being perpetrated against oppressed human beings. Like the unbridled destruction of the last wild places on the planet, these agonies are the combined outcome of selfishness and the misapplication of technology. While we are the 'Lucky Country', we are also part of the planet. We must tangle with the wider world problems or perish because of them. Inevitably, a test of our mettle in facing the greater challenges is to face those here in Australia, in our own backyard.

With Australia's wild face of two hundred years ago almost gone and the destruction proceeding apace, it is indeed a challenge to see how little is left and to save it from men and machines in search of instant extractable resources, profits and power. Fly over Australia and see. Few wedgetail eagles which soar high on the thermals today can look down on a territory unscathed by roads, pipelines, powerlines, buildings, furrows or fallen forests.

Australia, unlike the United States, has no true National Parks system – our parks are run by and are at the mercy of State Governments. Australia has no Wilderness Act (the U.S. Act was passed in 1964) or Wild Rivers Act (U.S. 1968). The Americans have the concept of and protection measures for wilderness enshrined in their law. We do not. We are starting so late, from scratch.

For wilderness to be saved, it must be understood. It is a large tract of entirely natural country where a person can wander, free of the distractions of modern technology. Free of roads, rubbish and the roar of machinery. It is nature on the broad scale. A place where people can see, hear, smell and feel the presence of nothing but nature and themselves. Maybe its single, most distinguishing factor is remoteness – from the roads, rubbish and roar.

Parks, rural backblocks and picnic spots are vital parts of Australia's heritage, but much more than these, wilderness is rare and special. It affords our souls the richest natural experience available in life. As the crush of civilisation grows in the years ahead, so wilderness will be at a greater premium as a refuge for the human spirit.

We are, after all, part of nature ourselves. We come from wilderness. Except for a few dozen generations at most, all of our numberless ancestors were children, women and men in wilderness. Wherever our origins, whether we be black or white, we were designed for life in the wild. Therefore, there is no-one on the planet who does not have an innate bond with wilderness. There is not one of us who cannot feel a special refreshment from kneeling to drink at a moss-lined mountain stream; who is not relaxed by the wash of waves on a secluded sunlit shore; who will not feel specially warmed by a bush sunset or grateful for the awakening which comes in the hush of a desert dawn. Not one of us can fail to feel intrigued, with all senses enlivened, by the stir of unseen creatures in the black of a rainforest night.

Now Australia is being stirred by the call to save the last of the wilds. Indeed, the loss of Lake Pedder in Tasmania's South-West wilderness in 1972-73 made a profound impact on the nation. This indescribably beautiful glacial lake, set amongst ancient rugged ranges, was flooded for a hydroelectric scheme despite international outcry. There were reasonable alternatives to the scheme, but the men in power would not listen or, at least, could not hear. In one of the world's worst ecological tragedies this century, Lake Pedder, with small unique aquatic creatures and its huge highland beach of pink quartzite sands, was drowned. The vast new hydroelectric impoundment is surrounded by those beautiful mountains, but also powerlines, quarries, scraggly roadsides, and neat, signposted rubbish bins. It is a testimony to the cold hard cheek of our materialist, technological age.

In the decade following the Lake Pedder furore, there has been an unprecedented advance toward saving Australia's wild places and species. Fraser Island, the Great Barrier Reef and the Colo wilderness near Sydney have been protected. Australia's last whaling station, at Albany, was closed and the near-extinct Lord Howe Island woodhen is being carefully nurtured. Yet the push for plunder has also gained momentum.

Uranium mines have sprouted in the wild Kakadu country of the Northern Territory; the rainforests of coastal Queensland are invaded by real estate signs, and the unique jarrah forests of Western Australia are being levelled.

In the face of this mercinary and merciless push, people are making a stand. In 1982-83, more than 1400 Australians were arrested and many hundreds jailed for peacefully protesting against the Hydro-Electric Commission's bulldozers and chainsaws at work in the forests by the Gordon and Franklin Rivers in Tasmania's South-West. That stand has drawn world attention to the needless destruction of a World Heritage wilderness. While reasonable options for generating electricity abound, rare wild rivers are irreplaceable.

The folk who are troubled by the turbulence of the protests in defence of the Tasmanian wilderness, should reflect on the uproar, jailings and tirades of official and respected abuse hurled against the suffragettes, the slavery abolitionists, or those people who fought to have children taken out of workhouses and given the right to a basic education. Each time humanity reaches for a better world, there is virulent opposition from those who hold political power or the thickest financial reins. Note that this truth is not confined to any particular political persuasion or part of the world community.

So the destiny of Australia's last wild places is in our hands. And it will not be easy. Though we are starting so late, the first step to saving our wild heritage is knowing about it. This book is about wild and wonderful places which typify the diverse natural heritage of our nation. It brings home the extraordinary beauty of some of the country's last wild places – places which most Australians do not yet know exist. It must stir the hardest heart.

The book is a call to us all to help ensure that such fragile natural grandeur survives through these years of worldwide resource extraction as a perpetual resource for human inspiration, recreation and reflection on the wild planet which used to be.

The book, its beautiful pictures and descriptions of wild places in each State of Australia, will gladden every heart. That such wild and wonderful places still exist, must stir the wildness, however dormant, in every reader. In that wildness, set free, is the key to saving the wilderness.

Dr Bob Brown
Director
Tasmanian Wilderness Society

Introduction

MOST JOURNEYS of significance begin with a goal, which can be as modest as that undertaken by author Bruce Chatwin when he set off for South America in search of the source of a tiny piece of fur sent to him as a child by an eccentric wandering uncle. The patch of skin had been snap-frozen on some prehistoric hide during the last Ice Age and the carcass had finally been dumped on the bleak shores of the southern oceans by a glacier. The search resulted in his stunning travel book *In Patagonia*.

Our goal was Australia's wilderness – an apparently easier task. It was not. When planning our 30 000 kilometre journey through all States, we were confident that our vast island continent would produce ample examples of beautiful, wild places for this book. Instead, our journey became a demanding search. It was a bitter lesson, but it ended on a note of triumph.

This big land has been 'tamed'. The proud, open spaces of the Outback, the monsoonal forests of Queensland and the Northern Territory, the mammoth dry wastes of Western Australia, the red-tufted dunes of Central Australia, the giant forests and alpine peaks of Victoria – our natural heritage – no longer exist as pure wilderness. In less than two hundred years of white settlement it has all been tainted with the legacy of intensive logging, grazing, mining exploration, weed introduction and land-clearing. Our wilderness is under siege.

Our journey began in the abundant rainforests of Washpool in northern New South Wales. Here were lush forests draped with lianas, sparkling wild streams and a rich variety of fauna and flora. Washpool, fortunately saved from destruction by a long conservation battle, is a pocket of wilderness to remind future generations of the towering rainforests that once carpeted the eastern coast of New South Wales and Queensland.

Off the north Queensland coast lies the jagged profile of Hinchinbrook Island. Here the human impact has been minimal because of the sheer ruggedness of the terrain, and huge forests and sprawling mangrove swamps paint its peaks and bays a rich green. A host of diverse and persistent tiny stinging insects add further protection. The Northern Territory encompasses the phenomenal diversity of the Kakadu National Park, but here, even rugged terrain and clouds of flesh-stabbing insects have not prevented the uranium miners from staking out chunks of the park.

This land-grabbing in the interests of mining in the Northern Territory, however, fades into insignificance when compared with the greed of the miners of Western Australia. Here, in Australia's largest State, there is the smallest percentage of land devoted to the national natural heritage. But, by some accident, one of this country's purest pockets of wilderness survives in the heart of the Fitzgerald River National Park in the south. Here, one of the world's richest plant habitats thrives. However, the pressures to absorb more park land for farming combined with the modern influx of four-wheel drive vehicles constantly threatens this unique biosphere.

We continued eastwards into the heart of Australia's biggest myth – the Outback. To our mainly urban population it may appear to be a vast untouched space, but in fact it is a huge scarred landscape criss-crossed with seismic exploration tracks and freckled with bulldozed airstrips. Luckily when we were there, the first Big Wet since 1974 made the Simpson Desert blossom briefly and temporarily hid many of the scars.

The low point of our journey was the hunt for the untouched landscapes of Victoria. Here, in Australia's most densely populated State, the pressures of logging and grazing have left only the smallest pockets of wilderness – and then only in the bleakest mountain gorges and gullies. From the air, the folds of the eastern half of Victoria are covered with what appears to be dense forest. The reality is that the canopy hides the legacy of 150 years of constant grazing. The floor of the forest has been stripped of new tree growth and the vast canopy of green hides a frightening problem for future generations. It is a forest of geriatric trees with no offspring, no regenerative powers.

Our journey ended in Tasmania. And here, in our smallest State, we found an example of pioneering 'foresight' – this beautiful island has Australia's largest percentage of National Parks and conservation areas. The rugged landscape has no doubt contributed to this 'charitable largesse' but unfortunately the wilderness battlefront is being waged there today. However, our journey through the wild frontier that is South-West Tasmania was a triumph, for here we found the most stunning example of wilderness – an unspoilt corner of this vast continent that hopefully future generations will be able to enjoy.

These precious lands – Australia's wilderness areas – are all under siege. They are all part of an internationally diminishing natural landscape. This book is a celebration of those rare, pure and wild areas that remain. Let's keep them free.

Allan Moult
Sydney, April 1983

A wild river flows free in the Washpool catchment area.
FOLLOWING PAGES: *Its brisk and refreshing journey continues.*

THE ABUNDANT RAINFOREST

Washpool, New South Wales

Dawn's mists soften the sun's early morning light in the Blackbutt Forest above Eaglehawk Creek where grass trees (ABOVE) *flourish. Temperate forest giants* (LEFT) *soar skywards.*

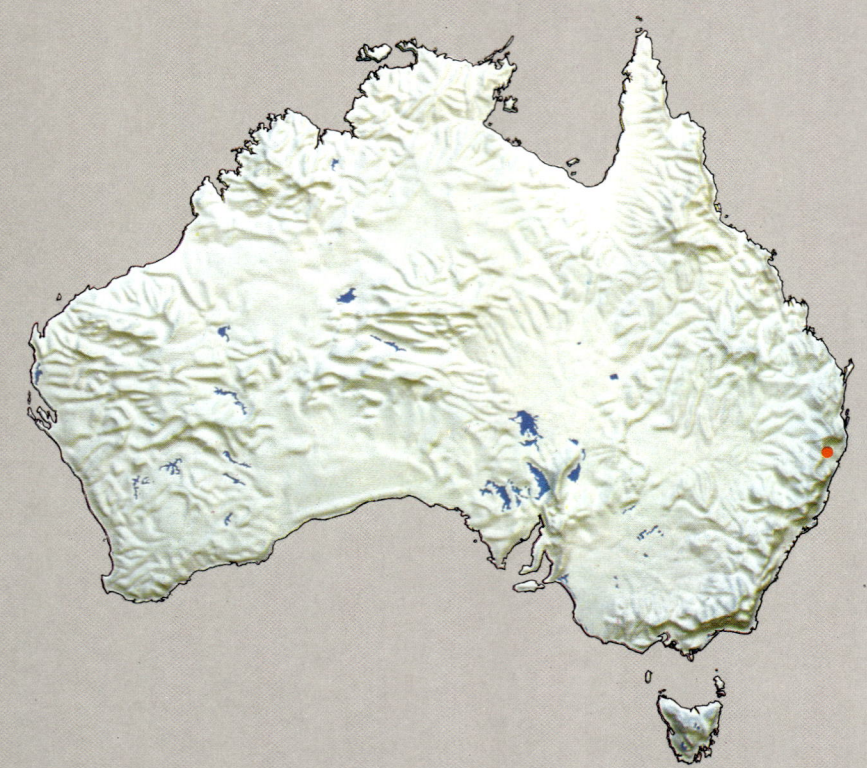

LOCATION: *Northern New South Wales.*
SIZE: *About 26 000 hectares.*
GEOLOGY: *Of mainly volcanic origin.*
CLIMATE: *High annual rainfall and generally cool climate.*
FLORA: *Contains the largest remaining unlogged and undisturbed temperate rainforest in New South Wales, with coachwood a dominant species.*
FAUNA: *Outstanding diversity because of essentially undisturbed communities, and a number of rare birds.*

WASHPOOL, NEW SOUTH WALES

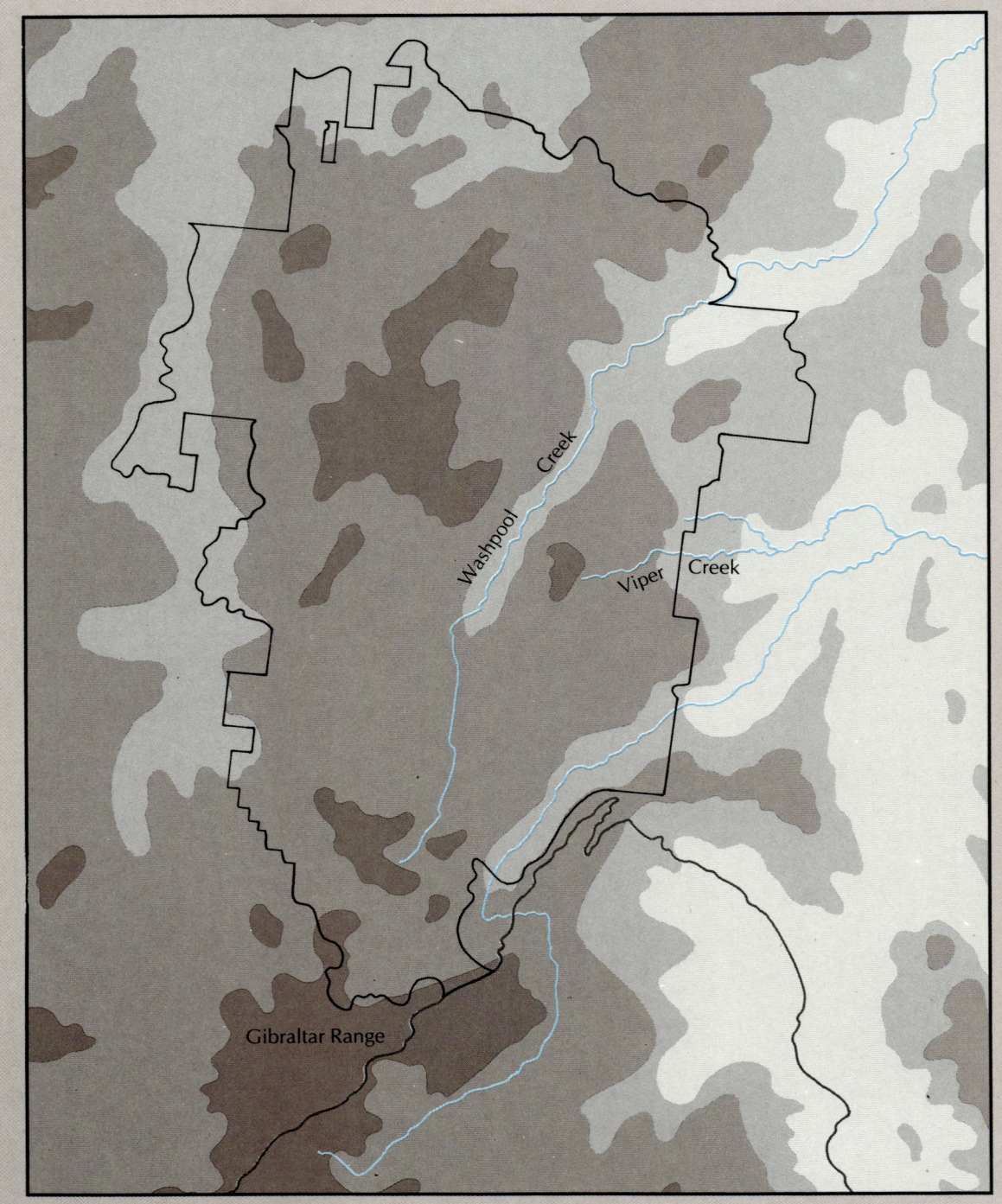

AN HOUR AFTER sunset the full moon begins to claw its way across the night sky. The rainforest canopy is a fluffy luminous sheet 50 metres above the forest floor litter. The moon's chill glow barely penetrates the ceiling of leaves. It is wind still. The turbulent gurgle of Washpool Creek, snaking through a sheer-sided gully 600 metres below Coombadjha Ridge, is totally muted by the bulk of the forest. To the east, the many tiny tributaries of Viper Creek are also silenced. The low and slow 'whoo-hooo' of a powerful owl mournfully hangs in the air. Suddenly, the forest comes alive.

There are tiny scurrying sounds throughout the layers of litter. Marsupial mice, tiger cats and a host of frogs busy themselves. A long-nosed bandicoot snuffles along; wave after wave of flying foxes swoop past with whispering wings; sugar gliders zoom swiftly through moonlit sclerophyll glades; bush rats and fawn-footed melomys feed nervously, their dainty paws grasping fallen seeds. A purple yabby postures with raised nippers as it hunts along the bank of a small swamp. The powerful owl hoots again.

In the gloom before dawn a heavy fog floats in the valleys. The forest is green and brown, dim and still. The eastern sky slowly lightens and the dawn chorus erupts. With an estimated 260 bird species in the area, it is a noisy affair. Among those contributing are the wonga, wompoo and topknot pigeons, the garrulous glossy black cockatoo, the rare double-eyed fig parrot and willie wagtails, whipbirds and pied currawongs. The chatter, chirping and squawking persist for nearly an hour.

Washpool Creek opens out into a wide pool.

Tall coachwood trees (ABOVE) dominate much of the temperate rainforest adjoining Viper Creek.
Large boulders frame a gurgling waterfall along Oorooroo Creek (RIGHT).

Freshly hatched yabbies (ABOVE LEFT) *form a glutinous mass as a proud parent brandishes angry claws* (ABOVE RIGHT).
Ferns thrive in the forest floor detritus (BELOW).

OPPOSITE: *Waves of lush forest flow to the misty horizon* (ABOVE). *A wary possum* (BELOW LEFT)
peers out between the branches and a big-eared bush rat (BELOW RIGHT) *seeks shelter under a mossy overhang.*

The Washpool area in far north-east New South Wales is part of a complex jigsaw of distinctive forest types which range from wet subtropical rainforest to dry sclerophyll woods. The diversity of tree species reflects this complex forest system. In close proximity can be found pigeonberry ash, yellow carabeen, coachwood, crabapple, Blue Mountains red mahogany, dwarf cypress, lemon-scented tea-tree, blue gum and white hazelwood. The environment also favours healthy stands of walking-stick palms, cabbage palms, hoop pines, palm lilies, elkhorn ferns, tree ferns and bird's nest ferns.

The dank rainforest sections also host dozens of climbing plants which compete vigorously for light high in the canopy. They are the prime colonisers of clearings or gaps created by storms or fallen trees and range from the large pepper vine and water vine lianas with their twisty, woody stems up to 25 centimetres in diameter and 60 metres long, to the delicate small vines such as the slender cucumber and forest bindweed barely three metres in length.

Moss-covered logs are strewn across the forest floor. They are micro-worlds of intense activity. Beetle larvae, amphipods and molluscs busy themselves chewing, scraping and digging in the constant process of decomposition.

Grass trees (RIGHT) *stand in rank beneath tall sassafras trees. Like green beards* (ABOVE RIGHT), *skeins of moss dangle from the forest's branches. A tumble of roots* (BELOW) *hosts a variety of plants.*

Coachwood trunks (ABOVE LEFT) splashed with lichen and giant eucalypts (ABOVE RIGHT) are the dominant species. Necklaces of moss and palm seeds (BELOW) are found everywhere in the rainforests. A giant liana vine hangs in twisted contortions from a host tree (RIGHT).

Sunlight, warmth and moisture are always abundantly present in rainforest and are stable and favourable throughout the year. They have been the same through long periods of geological time, and as a result, there is a tremendous variety of different kinds of organisms found and equally diverse relationships.

The thickest tangle of vegetation is the second growth that springs up after rainforest has been cleared or damaged by natural forces like cyclones. Small mammals and rodents thrive and multiply in this fringe vegetation, providing abundant food for the snakes that move in. It also provides refuge for mites, ticks, flies and mosquitoes. This is the 'green hell' of fable and fiction.

The luxuriance associated with rainforest has often lured farmers, who have destroyed the forest and reduced it to firewood. But their systematic clearings planted to pasture or crops have often failed. The lesson is rarely learned. The lushness of the felled forest might appear to offer high soil fertility, but the favourable moisture and temperature conditions primarily responsible for the vegetative bulk and complexity of the forest are also prime causes of advanced weathering and leaching of the underlying substrate.

The rotting mulch of the forest floor houses a thriving community of semi-slugs (BOTTOM LEFT AND OPPOSITE BELOW), beetles (BELOW) and other insects which, through their constant gnawing and chomping of the detritus, help break it down into a fertile mulch. Plate fungi (OPPOSITE) jut out from the base of a forest giant like shop awnings and in the intense competition of this rich and varied environment, many seeds (BOTTOM RIGHT) enjoy gaudy colours to attract seed-eating birds.

True rainforest – untouched, virgin rainforest – is very different. The forest floor is open and layered with the richly variegated browns of fallen leaves. This cover is easily scuffed away. There is no thick accumulation of leaf mould like that of deciduous forests; no rich accumulation of humus. The process of decay is too swift to permit much organic accumulation in the soil.

There is little vegetation on the forest floor because the light is far too dim for plants. A thin growth of tree seedlings may be waiting for a forest giant to fall and clear a space in the canopy for sunlight to flood through. Ferns, dwarf palms and scattered thick-leaved aroids, the umbrella trees that are favoured indoor plants, are also in the queue. The forest is a minefield of fallen trunks, tangled thickets, stretches of swamp and countless streams. The tall, solid, pencil-straight trees taper off high above in the vaulted green canopy they support.

There is an aura of calm. The cool, dim light, the utter stillness, the massive grandeur of the ancient red cedar, coachwood and buttress-based eucalypts, the nets of thick, woody lianas wrapped around tree trunks or looping down from the canopy above all contribute.

A mammoth coachwood tree and a giant smooth-trunked eucalypt (BELOW) share centre stage in the morning mist.

OPPOSITE: *The dense forest canopy (ABOVE LEFT) hides the elongated trunks of the many trees competing for a patch of sunlight (ABOVE RIGHT AND BELOW).*

It is easy to be overwhelmed by the physical form of a wilderness area as compact and variable as Washpool and perhaps one should heed the words of P. W. Richards in his book, *The Tropical Rainforest,* in which he claims: '[rainforest] vegetation has a fatal tendency to produce rhetorical exuberance in those who describe it.' But it is an amazing and exuberant world. The epiphytes – lichens and mosses, and ferns, orchids and bromeliads – form a kind of micro-desert as they stubbornly cling to dry branches high in trees. Elkhorn and bird's nest ferns catch water in their leaves and also feed off rotting organic matter trapped in their roots.

The epiphytes get the mineral salts they need either from the dilute solutions in the rain that washes them, or from the humus that collects in the bark of the host tree, or in the tangle of their own roots. Often fungi live in close symbiotic association with the epiphytes and help one another with food collection. The aerial root systems of epiphytes are often also used as nesting sites by ants and it is thought that the epiphytes gain an additional source of nutrition from the constant fossicking and haulage of foodstuffs ferried up to them by the ants. The ants in turn get a well-protected home built by the plant and provide food as a form of 'rent'. As many of these tree-based ants sting fiercely, they also provide protection for the epiphytes.

In rainforest areas there is a phenomenal number of different kinds of insects. The biggest known insects are found in this habitat – butterflies and moths with saucer-size wingspreads, giant soft-shelled cockroaches and elongated stick insects. The flies are big and the wasps also.

The fungi that are found everywhere in rainforests play a vital role in the breakdown of organic matter, releasing nutrients for re-use by the living plants. The buttressed base (OPPOSITE – TOP LEFT) is a common feature of the larger tree species and the epiphytic ferns – stag horns and elkhorns (TOP CENTRE AND RIGHT) – are often found high up in the forest canopy.

Of all forms of vegetation habitats, it is in rainforests that the 'the law of the jungle' has most meaning. The struggle for survival is a cold-blooded on-going battle. Here are found the most fanciful contrivances for catching food or for avoiding being caught for food – the most perfect examples of camouflage. The jungle law is symbolised by the strangler fig, which starts life as an epiphyte, a seedling growing high on a tree, sending down roots which reach the ground and grow until finally the host tree is smothered by the encircling fig, which then stands alone. The host tree finally rots away, leaving the strangler with a hollow, lacy trunk that is then invaded by spiders, birds and other creatures seeking secure nesting sites.

There are insect forms still found today that hardly differ from the fossils of their ancestors that lived 300 million years ago in the forests of the Carboniferous period. Typical of these are the numerous soft-shelled cockroaches found scuttling through the forest detritus. Under a rotting log there are usually found peripatus – a delicate, brown, multi-legged, caterpillar-like creature that is the ancestor of all the land arthropods – the millipedes, centipedes, spiders and insects. This anachronistic creature is surviving well in the warm, damp world of the rainforest, unchanged by the shifting hazards in hundreds of millions of years of forest life.

The rainforests appear to have more than their share of survivors from the past. Marston Bates in his fine book, *The Forest and the Sea,* remarks on this legacy and concludes: 'One can visualise the struggle for survival, the competition, the strenuousness of life in the rainforest. But then in the next instant, looking at this multitudinous accumulation of organisms one gets the feeling that there is so much warmth, so much light, so much moisture, so much food, that almost anything can survive and that almost everything does.'

We are reminded of the tiny doily-sized spiderwebs strung everywhere by long strands between the foliage. It is as if the spider, in spinning its trap, has considered the abundant insect life and decided not to work overtime in harvesting its catch.

The Washpool rainforests are a haven for a spectacular diversity of reptilian and insect life, including the mating grasshoppers and lizards (BELOW), the sinister wolf spider (BOTTOM LEFT) and the handsome ground-dwelling Nicodamus bicolor (BOTTOM RIGHT).

OPPOSITE: A lack of calcium in the detritus of the forests has evolved soft-shelled and shell-less semi-slugs (TOP LEFT), and a green caterpillar displays its false eyes which are meant to deter predators (TOP RIGHT). A strangler fig (BELOW) finally stands alone after choking its host tree, which has rotted away to leave the filigree network of the interlocked root system of the fig.

Inveterate traveller and explorer, W. H. Hudson, found rainforest almost overwhelming. In his book, *Green Mansions,* he wrote: 'Here nature is unapproachable with her green, airy canopy, a sun-impregnated cloud – cloud above cloud – and though the highest may be unreached by the eye, the beams yet filter through, illuming the wide spaces beneath – chamber succeeded by chamber, each with its own special lights and shadows.'

His impressions of Washpool would perhaps have extended into the realm of 'rhetorical exuberance' once confronted with the variable mosaic of forest types created by the extremes of geographical features – the steep gorges sinking a vertical kilometre into the green seas fringing Washpool Creek, the volcanic ridges crowned with 70 metre blue gums and the wild rivers that course jubilantly through the wilderness.

The streams and creeks that fan out through Washpool are sparkling reminders of the days when vast luxuriant rainforests stretched intermittently along the entire New South Wales coastline. Today less than half survive. The remainder, with isolated exceptions already incorporated into National Parks, are under threat from timber companies or land-hungry farmers. Perhaps more men of the calibre of John Lever are needed. Over one hundred years ago when he was given the concession to log the rainforests on Lever's Plateau in north-eastern New South Wales, he decided the forests were too beautiful to cut and there they remain, untouched to this day.

The quiet pleasures of the rainforest environment reveal themselves in the spotlit canopy (LEFT) *and in the leafy gullies with their bubbling creeks* (BELOW AND RIGHT).

Early evening. There is no real twilight in these heavy forests. As the sun sets, its horizontal rays are absorbed by the dense foliage and shadows deepen rapidly. The creatures of the day – mainly birds and reptiles – settle down before the awakening of the night shift. The bats are the first of the nocturnal forest denizens to become active. The powerful owl broadcasts its mournful dirge.

The frogs begin their incessant croaking. Among them is the curious marsupial frog, *Assa darlingtoni*. The male of the species, a broad, squat creature, has taken on the role of looking after the offspring and carries the developing tadpoles in pockets in its groin. Females lay up to ten large eggs and the male then sits in the middle of the spawn mass when the tadpoles hatch and waits for them to locate the opening of a sac. They then force their way inside, using their heads to push open the flap in the skin. As they grow, the tadpoles compress their father's stomach and other vital organs so much that he finds it impossible to feed. By the time they set off on their own, they have grown to half the length of their 20 mm long parent – stretching the pouches to almost the entire length of the male's body.

The green tree frog, *Litoria caerulea,* perhaps the best known of all Australian frogs, is also resident in Washpool. Of a most amenable disposition, it is remarkably tolerant to being handled by humans, and one we discovered disporting itself near Viper Creek proved no exception. Picked up gently, it simply snuggled down to make itself comfortable on the palm of the hand.

A scrawny brush turkey (LEFT) *forages in the undergrowth, and a long-toed frog* (BELOW) *clambers to safety.*
OPPOSITE: *A New Holland honeyeater* (ABOVE) *feeds its chicks and a yellow-faced honeyeater* (BELOW) *lands to feed its own.*

Wilderness as a description of habitat is usually confined to the human scale, yet in the micro-world of the insects of the rainforest, there are wilderness areas too. It is a surprisingly crowded and violent world.

The diversity and abundance of insects is a result of the great number of plant species found in the area. Humidity and high temperatures also favour the fertility. For example, of Australia's 364 species of butterflies, 305 occur only in rainforest. This has led to incredible population statistics. The primitive 2-3 millimetre long wingless insects known as springtails, whose antecedents have been found in fossil form in Devonian shales, number up to 60 000 specimens per square metre. After rain, they are often washed into pools where their sheer numbers create grey-blue scums. Besides fouling the water, they play an important part in the food chains of the higher predatory insects. This food chain includes tiny millipedes, minute beetle larvae and bract and gill fungi, which in turn attract bigger fly and beetle larvae. Some fungi are luminous and their light gives an X-ray effect to the busily feeding transparent insects.

The larger beetles include the Passalids, a gregarious mob which live together in family groups in rotten logs. Researchers believe they form pair bonds for life and they seem to show a primitive social organisation. Their larvae have a large, dilated hind gut, which works as a fermentation chamber where the cellulose in the decaying wood they feed on is broken down by symbiotic bacteria and protozoa living in mutualism.

A related species, the carab, is the terror of the condensed world of insects. They are active nocturnal foragers for earthworms and snails, and when pursued or threatened they eject a pungent acidic fluid which forms a small cloud of smoke-like gas to cover their escape.

The fungi of the forest assume a spectacular variety of shapes, colours and textures as they break down the forest floor detritus. The mosses (ABOVE RIGHT) are also a common feature of the forest.

At our Coombadjha Ridge camp site on the last day we listen to the forest birds' dawn chorus subside from the raucous celebratory greeting of a new day to the mundane chirps and cackles of everyday scavenging, feeding and flirting. The fog that closed in overnight has dissipated. Tiny, piercing patches of sunshine freckle the forest floor.

The sun's path gradually unveils a two-metre snakeskin entwined in a clump of undergrowth. It is still pliable warm and musty. The head is perfectly formed, even one eye lens is still attached like a droplet of water. There is a slight movement a few metres away and a dark shape emerges from the shadows. The snake elegantly curls up in the sun, black back gleaming like patent-leather, red belly glowing like neon. Exhausted from the early morning effort of shedding its skin, the red-bellied black snake slowly relaxes and goes to sleep. We tiptoe away.

Our journey through Australia's wilderness has begun, yet in a way, Washpool seems as much a final triumph as a beginning.

Rich growth characterises the temperate rainforest of Washpool (RIGHT). *An appleberry* (BELOW LEFT) *retains its bell shape after falling from the canopy high above. A waxy orchid* (BELOW RIGHT) *contrasts with a spiky tree flower* (BOTTOM LEFT) *and the red fruit of a bangalow palm* (BOTTOM CENTRE). *A lone appleberry* (BOTTOM RIGHT) *sways in the wind.*

Spiky-leaved casuarinas (ABOVE) *at Macushla Bay frame Mount Bowen and its sister peaks across Missionary Bay.*
FOLLOWING PAGES: *Pre-monsoon clouds build up in the late afternoon as Hinchinbrook Island floats in a mirror-like sea.*

THE MANGROVE OASIS

Hinchinbrook Island National Park, Queensland

A sunset in a tropical paradise. A setting sun at low tide silhouettes young mangrove trees on the beach.

LOCATION: *160 km north of Townsville.*
SIZE: *68 400 hectares.*
GEOLOGY: *Volcanic origin with mainly granite soils.*
CLIMATE: *Monsoonal with up to 8000 mm rainfall each season.*
FLORA: *Coastal regions are predominantly mangrove swamps; mountain areas are basically eucalyptus forests intermingled with pockets of tropical rainforest in many valleys.*
FAUNA: *Rich and diverse, especially in varieties and species of fish, crustaceans and avifauna.*

HINCHINBROOK ISLAND NATIONAL PARK, QUEENSLAND

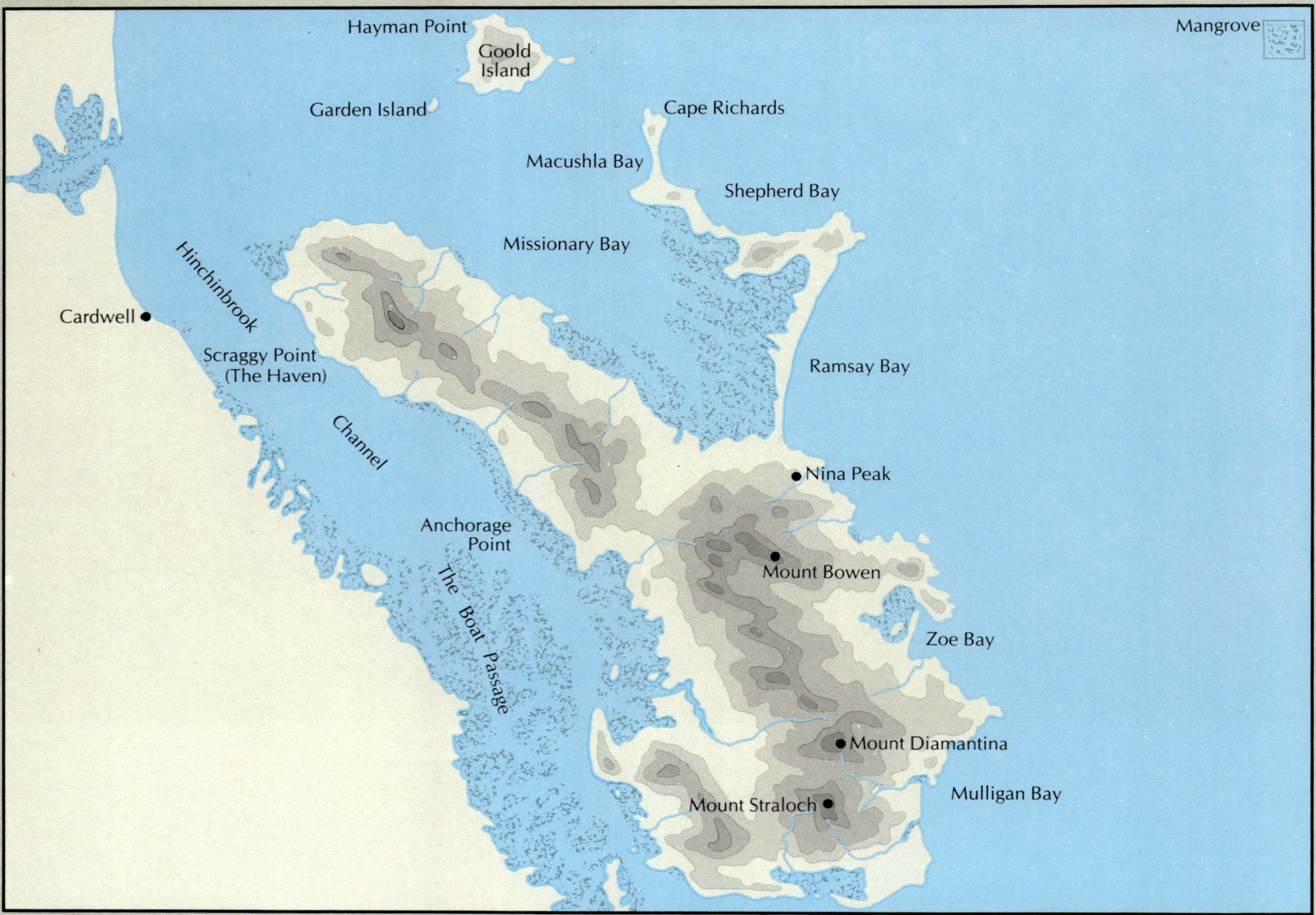

Mangrove

Hayman Point

Goold Island

Garden Island

Cape Richards

Macushla Bay

Shepherd Bay

Missionary Bay

Hinchinbrook

Cardwell

Scraggy Point
(The Haven)

Ramsay Bay

Channel

Nina Peak

Anchorage Point

The Boat Passage

Mount Bowen

Zoe Bay

Mount Diamantina

Mulligan Bay

Mount Straloch

BOVE, THE SOLID CANOPY of the mangrove swamp absorbed the brunt of the heavy monsoon rains that had been falling all day. Its chunky leaves channelled the walls of water along branches and trunks like downspouts. The noise was awesome, but still did not drown the belches that rose from the belly of the dank, dark mud below. It was a primeval world, a bizarre movie set filled with the multi-rooted mangroves – a world of deep shadows, no breezes and odd splashes produced by strange creatures.

To some it is evil. Yet, this weird landscape of soft, oozy black mud, haunt of sandflies and mosquitoes, and twilight haven for solid crabs sheathed in armoured carapaces is also home to the mangrove honeyeater, recognised among birdwatchers as perhaps the sweetest singing honeyeater of all. And its loud and rollicking song, 'go-bidger-roo, go-bidger-roo', like a snap of the fingers, puts this strange world into perspective – it is just one more biological nursery, and with little doubt one of the most essential in the complex food chain of the oceans of the world.

A buttress-rooted mangrove with a tortured trunk stands aloof in a sea of black ooze (RIGHT). Bright green mangrove trees strut on suspended roots in one of the many channels that criss-cross Missionary Bay (BELOW).

Hinchinbrook Island squats in mountainous splendour a few kilometres off the northern Queensland coast. From a boat in the Coral Sea it appears to be part of the mainland, which it was before the last Ice Age retreated, and to this day it shares many life forms in common with the Australian continent. It was originally part of an eastern mainland mountain range which paralleled the Great Dividing Range. Today it is Australia's largest island national park (68 000 hectares) and is custodian of vast areas of untainted mangrove swamps.

Many organisms from minute bacteria and primitive fungi to clams, crabs and oysters are dependent on mangroves for food and shelter, and the subtleties of the interactions between them is being intensively studied. The complexity of the life forms in this area of transition from the sea to the land has fascinated scientists for centuries as it is an obvious route from sea to land in the evolutionary process.

The immense importance of this food chain in the ocean's lifecycles is evident at the most basic level – that of mangrove bacteria. They are an important component in the diet of prawns and mullet, which in turn play a major role in the food webs of larger marine creatures. This is also true of the algae, fungi and lichens in the swamps.

Mangrove swamps (ABOVE) *dominate the waters of the Hinchinbrook Channel and carpet the broad sweep of Missionary Bay* (RIGHT).

Low tide (ABOVE AND RIGHT) *reveals the sculptured complexity of these buttress-rooted mangroves, hidden deep in the swamps of Hinchinbrook Island.*

Tree trunks and roots contort into strange configurations in their effort to establish a firm footing in the swamps and rainforests.

The roles are many. The minute protozoa spend their days decomposing mangrove seedlings and other detritus, and in the process often fall prey to the box jellyfish. Many species of worms thrive in the mud. Oysters and molluscs cling to the mangrove roots and at least 52 species of crustaceans, especially micro forms and larvae, are widespread. Spiders, midges, mosquitoes and beetles bite and get bitten . . .

The mangrove estuaries and creeks are favoured nurseries for many juvenile fish and 68 species of fish (excluding sharks and rays) have been recorded in these shallow waters. They are also the home of the bizarre mudskipper, which skips about on the roots and mud in and out of the water. It is the architect of the turreted burrows so common in mangroves.

The list grows. Both land and sea snakes are found, as are lizards, geckos, skinks and turtles. And there is *Crocodylus porosus*, the saltwater crocodile, needlessly the most feared indigene. (Recently a four-metre specimen was shot a few kilometres north of Hinchinbrook Island.) Intensive hunting in the area late last century decimated the reptile, but in recent years they have made a gradual comeback. Shy, territorial creatures, they mainly feed on crabs, prawns and fish.

Moss-covered boulders cracked by weather shelter among the spindly growth in the pocket of rainforest on the shores of Shepherd Bay.

A cool green rainforest canopy filters the sun and brings a permanent twilight to the forest floor.

Every nook and cranny of the rainforest is sheltered by overhanging fern fronds and palms (ABOVE AND BELOW).
OPPOSITE: *Clockwise from top left. Subtle shades colour the forest debris; a mushroom's mottled cap traps rainwater; a climbing fern clings to its host and a python-like liana strangles a tree.*

A vine noose (ABOVE LEFT) *hangs tight on a palm trunk. A parasite takes advantage of a fire-damaged tree trunk* (BELOW LEFT). *A large paperbark eucalypt contributes to the forest detritus* (ABOVE).

The mammals of the mangroves include fruit bats and flying foxes, water-rats and swamp wallabies. Naturally this wealth of animal life has attracted many birds and according to recent studies there are an estimated 250 species.

Yet for all its vibrant and diversified lifeforms, the mangroves remain a mysterious domain. From the slopes of Mount Bowen, the mangrove swamps that cover Missionary Bay look like a green scum dissected by meandering strips of water. At low tide the sea recedes to reveal huge mudflats striated with swift-flowing, frothy tidal creeks – an unwelcome and messy environment at best. Closer study, however, reveals an entrancing world, a complex biological Eden as delightful as the call of the mangrove honeyeater.

At twilight the ghost crabs scuttle down to the tide's edge for an evening concert as old as time. Pirouetting and prancing on fine-pointed ballet toes they leave lacework tracks in the wet sand. Their busy and apparently aimless scurrying is a search for food carried ashore by the tides. Their frantic movements provide protection against scavenging birds and their own kind. In the cannibalistic world of crabs, it pays to be alert.

Their translucent, sand-coloured bodies flecked with pink and yellow provide perfect camouflage and they add to their safety armour by being swift and efficient burrowers. Home is a hole in the dry sand far above high watermark, a hole up to a metre deep. Twilight sees them scurrying along the frothy edge of a dying wave seeking titbits. Success brings them to a dead stop and the wave keeps going, leaving only a pair of pop-eyes on periscope stalks above the surface.

A confident Australian pelican (ABOVE) *comes in for a smooth landing on a sandbar shared by a flock of young birds* (BELOW).

OPPOSITE: *Birdlife thrives on the fringe of the mangrove swamps. The mangrove detritus* (BELOW) *may look like scum but it is a biomass full of edibles for the crabs and other crustaceans.*

A large egret (ABOVE) swoops over the island, while a timid zebra pigeon (BELOW LEFT) and an immature sea eagle (BELOW RIGHT) hide in the branches of the forest.

OPPOSITE: *A pioneer mangrove stranded off shore by a low spring tide creates a surreal setting for the distant bulk of Hinchinbrook Island.*

On the sandy flats to the west of Missionary Bay an ebbing tide brings the blue-mauve soldier crabs out onto the sandflats, little bodies high on stilt-like legs. In tight groups they forage slowly, their claws constantly picking up edibles from the surface layer of the shore and transferring it to their busy mouth appendages. Observing their progress calls for patience. They are skittish, nervous creatures and any sudden movement sees the hordes scatter and disappear by burrowing into the wet sand with a corkscrew motion.

Further south amid the tangled roots of the mangrove swamps, the semaphore crab signals its presence by raising the front of its body rhythmically with nippers outstretched in a come-hither motion. Typically nervous, the semaphore crab stays near its burrow, but in an emergency will dive down the nearest hole it finds.

Here too are the strange fiddler crabs with their solitary bright red or orange claws – often as large or larger than their bodies – raised pugilistically. Because of their conspicuous actions or colours, many crabs have been given simple descriptive names, but perhaps the most common – a nondescript muddy olive crab with long eyestalks which flap down sideways into a protecting groove if touched – has to be content with its unwieldly scientific nom de plume *Macrophthalmus crassipes*. When the tide is out these crabs can be seen in their hundreds crawling through the mud, particularly where the narrow-leafed zostera (dugong grass) is found growing. Perhaps the ultimate mud crab, it has had to forego even that humble title to the renowned and tasty *Scylla serrata* of the Queensland mangroves, the large crab that gourmets drool over, which unlike other crabs stays hidden in its burrow while the tide is out.

In deeper water, the hyperactive blue swimmer crab thrives. It is superbly adapted to its habitat. Its last pair of legs are shaped like paddles and its crisp, white belly and mottled, brilliant blue back give it protective colouration from air or ocean floor. It is also tasty and like the mud crab, it supports a thriving commercial fisheries industry.

Frothed tidal waters (BELOW) *provide a smorgasbord for crabs and birds.*
OPPOSITE: *A large goanna* (TOP) *gives an unconcerned backward glance. A battle-scarred but still pugnacious crab guards its burrow* (CENTRE LEFT). *Empty sea urchin shells* (CENTRE RIGHT) *often dot the beaches. Curious mudskippers eye-balling through the shallow waters* (BOTTOM).

Hinchinbrook Channel and Missionary Bay are browsing territory for the barramundi which has become victim of its reputation as Australia's tastiest fish and is now in rapid decline. This is happening despite it being one of the most fertile fish in the world. Research by the fisheries division of the CSIRO has shown a single female can produce up to 46 million eggs in one breeding season. The barramundi, however, has a curious sexual cycle. Halfway through its life it changes sex from male to female and this is a problem – small barramundi are male and big ones are female. It is a rare fisherman who throws the big ones back.

The narrow channel between the island and the mainland is a favoured haven for small craft when cyclones threaten. It is also a favoured feeding ground for the pelagic leathery turtle and the nomadic green, loggerhead and hawksbill turtles which visit from time to time. They all have the disturbing habit of suddenly breaking the surface for a quick stocktake of the world above, before swiftly disappearing.

Other visitors include schools of dolphin swimming with synchronised precision and the occasional humpback whale. Before 1950, these 14 metre marine mammals were a common sight in waters of the Great Barrier Reef as they passed by on their annual migration between the Antarctic waters and the tropics, where their calves were born. Then in the 1950s whaling stations were set up on the New South Wales and Queensland coasts, and together with the long-established Antarctic hunts by the Soviet Union and America they were slaughtered in their thousands. By the time the whaling stations on the eastern Australian coast closed in the early 1960s, it was estimated that only two hundred remained in these waters. Today, their numbers are slowly increasing, but sightings are still rare.

The red carpels (BELOW) are all that remain of this flower of the Golden Guinea tree (Dillenia alata). Its yellow petals have already fallen off in the morning sun and its white seeds have been eaten by parrots. An umbrella tree (ABOVE RIGHT) (Schefflera actinophylla) displays its distinctive fingers of red flowers. This bauhinia (Bauhinia galpinii) (RIGHT) is an introduced species which has adapted well to local conditions.

The flowers of this common seashore native hibiscus (Hibiscus tiliaceus) (ABOVE LEFT) are yellow when they first bloom, but gradually turn red after falling on the sand. Sterile fronds (ABOVE RIGHT) of crow's nest ferns (Drynaria rigidula) and the pale orange flower of Cordia subcordata (BELOW) are found just above high tide mark.
OPPOSITE: The thin fragrant flower of this native shrub, Clerodendrum inerme (ABOVE), contrasts with the splayed white petals of the sea lettuce or Cardwell cabbage tree (Scaevola taccada) (BELOW).

Brightly coloured beetles (ABOVE) scurry about investigating a seed
pod. A splendid example of the mimicry of nature – two empty shells
of a legume seed pod (BELOW) pose like the wings of a moth.
The short-horned grasshopper (RIGHT) is one of the most abundant
insects, but it readily falls prey to birds.

Our campsite is nestled behind a low sand dune on the narrow spit of land that separates Ramsay Bay from the mangrove swamps to the north. It is ideally situated for protection from the weather – perhaps too well. Nightfall sees us head for the crest of the dune to capture the faint breeze. Our camp has become a murky basin of fetid air populated by hundreds of kamikaze mosquitoes and bullying cane toads. Relief is short-lived. A new moon rising soon after midnight reveals black clouds rumbling on the island's precipitous peaks. Half an hour later the storm breaks bringing more than 100 mm of rain before dawn.

Traipsing along the 10 kilometre length of Ramsay Beach we are attacked by minute white-winged black sandflies. Back at camp we inadvertently disturb a green tree ant nest. This precipitates a full-scale attack and they add insult to injury by spraying their painful bites with formic acid exuded from their bellies. The reckless hordes continue their attack despite retaliatory blasts from aerosol cans. We retreat to the beach.

These formidable insects display unusual co-operative action in building their nests. Their communal abodes consist of leaves drawn together and bound with silk produced by their larvae. Worker ants form extensive chains with up to a hundred or more ants joining in to drag leaves to the nest site.

By nightfall they had settled down and we are treated to the muted fireworks display of *Lucio costata*, one of Australia's rare species of fireflies. The light is the result of oxygen being supplied to an enzyme, luciferase, which is located in the cells at the end of the abdomen. Eggs, larvae and adults glow, but the light produced by the adults is the brightest and is a mating signal.

A tidal flat behind the dunes of Ramsay Bay fills these distant views of Mount Bowen (ABOVE AND LEFT). Heavy monsoon rains drench the island's peaks (BELOW) while bright sun paints the lower forests.

Driftwood sculptures decorate this isolated beach at the foot of Nina Peak on the island's east coast.

OPPOSITE: *Mist wafts through the dramatic, sheer peaks that buttress Mount Bowen.*

The many rugged peaks of Hinchinbrook Island are often hidden by cloud. Their gaunt flanks are near-sheer cliff faces lining steep gullies and ravines. Below Mount Bowen, which dominates the central section of the island, a particularly scraggly peak stands out in stark relief against the misty clouds. Its corroded summit is a jumble of boulders separated by expansion cracks. One isolated mass takes on the shape of an eagle hunched in readiness for flight.

It is an island that has been barely tampered with by human hand. Early this century red cedar was logged in the more accessible valleys. A small pleasure resort trades on the northernmost point, Cape Richards, and controlled campsites exist at Scraggy Point, Ramsay Bay and Macushla Bay. An isolated coconut grove marks the original site of a Depression-era resort at The Haven which failed to attract a single client, even though it was offering a week's holiday for just one pound!

Today Hinchinbrook Island is still a unique wilderness of volcanic peaks, tropical rainforest, eucalypt forest and vast mangrove swamps. Most of the rainforests are confined to the western and southern lowlands and gullies, with quandong and fig trees, maple and myrtle dominating. A veil of vines and epiphytic ferns compete for sunlight below the towering canopies.

To the north, however, on the curious scythe-shaped finger of land that protects the mangrove swamps from the Coral Sea weather, a small hill presents a strange, dark-green profile against the paler hues of its eucalypt-covered neighbours. Here giant trees jut skyward their upper branches swathed in clinging, choking vines. A freak of geography has created an ideal environment for this isolated rainforest. On its gentle slopes the hill nurtures 20 metre high palms, giant paperbark trees, Moreton Bay ash and a host of parasitic plants. At its foot, wavy buttress-rooted mangroves – one of over thirty species found in the area – give way to other mangroves which test the water with delicate aerial roots. Their march to the sea's edge is a gradual transition with hardier species taking over until the pioneer zone is reached. Here *Avicennia marina* has established the beachhead for the tidal forests.

The root systems of the 30-odd species of mangroves found on Hinchinbrook Island are prolific, determined examples of exercises in survival.

The heavy monsoon rains rinse the skies and clean the beaches to give Hinchinbrook Island the facade of paradise. Beneath the forest canopies the mosquitoes thrive and on the beaches the persistent sand flies hold court.

Our visit coincides with boisterous Cyclone Elinor. The sand bank shelters us from the brunt of the gale, but by morning the full 10 kilometre length of Ramsay Beach has been restructured by the fury of the seas combined with a three-metre rise in tide. Millions of tonnes of sand have been lifted high up to rows of scraggly casuarinas that line the beach. Gone is the gently contoured sweep of white sand that met yesterday's seas. Now a steep wall of wet sand holds the choppy swell at bay. The tidal fringe is a morass of debris and driftwood alive with frantic crabs celebrating the windfall.

A few days later, the long-awaited monsoon abruptly arrives. Overnight, the island is transformed. Smelly creeks sluggish with debris and punctuated with wriggly mosquito larvae are flushed, limp-branched trees lose their hang-dog stance, grasses and shrubs sprout green shoots, vast schools of prawns congregate inshore, the murk of the mangrove swamps is filtered and the flanks of the many mountain peaks shimmer with rampant waterfalls. During the first four days, over 1015 mm of rain is recorded.

This is the lesson of Hinchinbrook Island. Its rugged landscape is often battered by nature's wilder forces, yet its swamps and rainforests with their delicate biospheres survive and prosper in the shadow of its protective embrace. It is true wilderness. There is balance.

As the suns sets behind the coastal range of mountains on the mainland, the island projects the calm of balanced wilderness.

The dawn of a new day brings glowing colour to the golden-hued sandstone rocks of the main escarpment in Kakadu National Park.

THE UNTAMED NORTH
Kakadu National Park, Northern Territory

The bleak expanse of the floodplain stretches to infinity (ABOVE) *from the Arnhem Land escarpment* (RIGHT).
FOLLOWING PAGES: *The fringe area links the floodplains and the escarpment.*

LOCATION: *Northern Territory.*
SIZE: *6144 square kilometres.*
GEOLOGY: *Sandstone, schist, quartzite and siltstones predominate.*
CLIMATE: *Cool and dry winter, hot wet summer with average 1200 mm rainfall.*
FLORA: *The diversity of landforms – floodplains, escarpments, coastal dunes and mangrove swamps – all contribute to an exceptionally rich plant life.*
FAUNA: *41 mammal, 250 bird, 75 reptile, 22 frogs, 45 fish and 10 000 insect species have been recorded.*

KAKADU NATIONAL PARK, NORTHERN TERRITORY

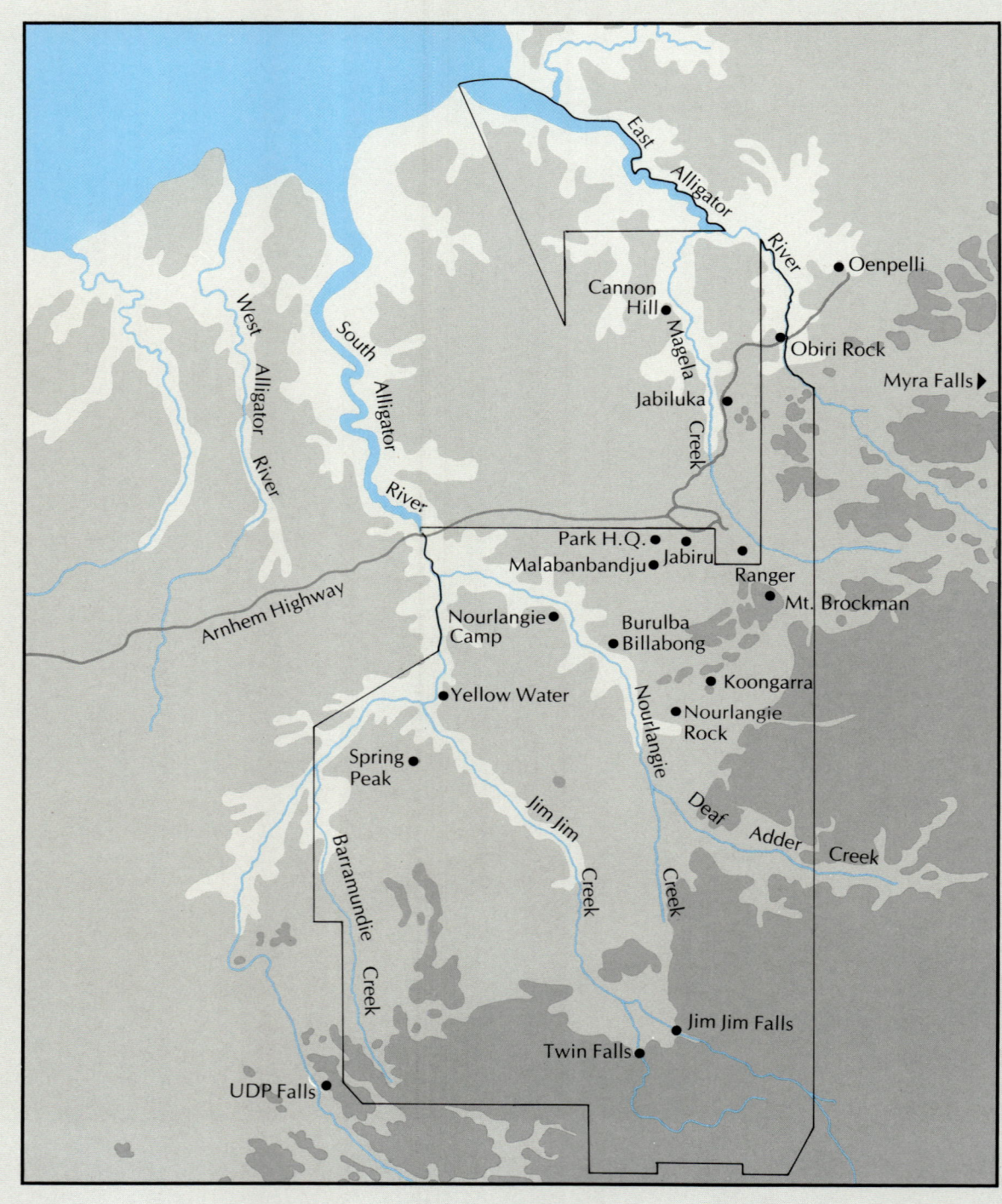

LIKE A SEA OF buffalo milk fresh from the udder, the morning mist froths as the first breezes tease it and reveal glimpses of the vast floodplain to the west of Obiri Rock. Totemic pandanus scrub wafts gently, spiky crowns aloof on their long skinny stems. An ungainly black-necked stork strutting across the mud scabs of a dried-out billabong punctuates the scene with an inverted 'V' as its long blue-black neck and equally long beak move above the misty landscape. The scrawny outlines of the scrub and the stork resemble the tableaux painted in ochre on the wall of a nearby rocky overhang. Here, Aboriginal artists have created a hunting scene with a series of stick figures. The primitive outlines have a curious grace and fluidity as with arms raised they prepare to throw their long stone-bladed spears. They have been holding that pose for at least 18 000 years.

The Arnhem Land plateau at the 'Top End' of the Northern Territory, is an 80 000 square kilometre sandstone massif that soars up to 250 metres above the adjoining lowlands. Its deeply serrated western fringe is an abrupt escarpment that slices through Kakadu National Park – a gift to the nation from the traditional Aboriginal owners. The park's birth pangs were traumatic. In the early 1970s interest in the uranium deposits in Arnhem Land finally prompted Justice R. W. Fox's Inquiry into Uranium Mining which was given the task of deciding whether uranium mining should be allowed in the Alligator Rivers region, whether it should be made into a national park, and whether the local Aborigines had a legal claim to traditional land ownership. In 1978 the Aborigines won their claim and were granted land rights to most of the Kakadu area between the Alligator Rivers. They had already been granted ownership of the Oenpelli Reserve, covering most of Arnhem Land to the east of Kakadu.

The setting sun peeks under a rocky overhang at Obiri Rock covered in Aboriginal paintings.

They then agreed to make Kakadu into a national park, as the area would be preserved in its natural state. However, after the Fox Inquiry, the Federal Government decided that major uranium developments should also be encouraged to go ahead and the Ranger Project Area on the northern border and the Koongarra Special Mineral Lease Application, 12 kilometres to the south, were excised from the proposed park area. In his conclusion to the Inquiry, Justice Fox had stated: 'In terms of human welfare, it is a question of accepting responsibility towards future generations, as well as for those now living. Imponderable questions arise which must largely be dealt with as a matter of faith; questions such as how finite are our resources and what trust can be put in science and technology to find answers to our problems. Ethical judgements are involved at many points and are the ultimate determinant.' The case rests.

The first Aborigines moved into the area at least 30000 years ago. Edge-ground stone axes, the oldest known in the world, were made here and a stone used for grinding ochre at least 18000 years ago has been found. This makes the stick figures at Obiri contemporary with the palaeolithic cave paintings of France. They also display a similar sense of fluid design and stylisation of animals. The Arnhem Land paintings, however, were still being renewed until recently – perhaps a generation ago – and the style of the 'X-ray' art lives on in current bark paintings.

The rugged grandeur of the Arnhem Land escarpment is revealed in all its time-scarred glory.
Aeons of erosion from wind, rain and sun have left its surface a pitted mass of broken rock.

Among the styles of painting that have been described are 'Mimi art', showing stick figures representing the 'Mimi spirits' engaged in battle and hunting; a very ancient style showing miniature figures, often in elaborate costumes; a larger figurative style; and the X-ray style, mainly used for animals, which shows internal organs as well as the animal's outline. The colours are mainly red and white ochre, and charcoal black. Some tribes used a blue pigment and others occasionally used blood.

The original artists probably painted while sheltering from the wet during summer and more than three hundred and eighty sites have been recorded in the region, while about one thousand more are known. Modern intrusions – mining, exploration, tourism, feral animals and introduced plants – are all threatening to destroy this priceless legacy, but their impact fades in comparison to the destructive talents of *Scelipgron laetum* – a large black-and-yellow mub dauber wasp.

Unfortunately, the overhanging rock shelters favoured over the centuries by Aboriginal artists are also favoured nesting sites for the wasps. They are shady, dry spots close to water, providing the mud that the wasps need to build their nests and that the artists, who worked with ochre-tinted mud, needed to mix their 'paints'. The strong, thick-walled mud nests built by the female wasps are extremely difficult to remove from the paintings without damaging the art beneath. CSIRO research has found that part of the problems may be caused by humans.

Wasps need mud to build their nests. Where once the water to make mud was available only during the rainy season, the insects now find water all year-round at new park facilities with leaking water tanks and dripping taps. Solutions to the problem include soaking the nests off with mild detergents when they soften in the wet season; using spray-on pesticide coatings to prevent nesting and even using delayed-action pesticides. Each has its limitation and it is now proposed to isolate the paintings behind fine wire screens, or even glass, until an effective long-term solution is found.

The invasion of the uranium miners (ABOVE) has left vast scars on this ancient landscape (BELOW). An Aboriginal rock painting in the 'Mimi art' style (LEFT) appears to be giving vent to the intrusions through wide-open lips.

The rocky overhangs of Obiri Rock (OPPOSITE BELOW) and nearby tumbles of eroded rock (OPPOSITE ABOVE) hold a priceless legacy of Aboriginal rock art (ABOVE) . Unfortunately, pollution and invasions of mud wasps (BELOW) are threatening their existence.

In the pastel light of the setting sun, a solitary ghost gum rooted in an escarpment fissure glows softly, its smooth white limbs absorbing the dying rays. It is aptly named, for as the tropical night suddenly descends, it still stands out against the blackness. Earlier, we had walked along dry streambeds and billabongs in the shade of broad-branched paperbark trees, our feet crunching the layer of dried leaves that gave the area the feel of a tame European forest. Beyond the water-courses, eucalypts, spinifex and wiry grasses dominate the plains.

The topography of Kakadu National Park has a dramatic profile from the Arnhem Land Plateau, through the lowlands and floodplains and on to the tidal flats. It is also combined with a changing soil balance. From the sandstone of the plateau it gradually changes through yellow-red, sandy and loamy soils to black acidic clays, saline grey soils to dune sand and saline mud. This diversity creates a mosaic of plant communities, including monsoon forests in small pockets near the coast, the semi-deciduous forests found in sheltered rocky sites and sandstone monsoon forests dominated by evergreens. In addition, there are savannah grass-lands, mangrove zones in the estuaries, heathlands on the sandstone flats, salt-tolerant fleshy samphire groups and sedgelands that survive under water for two to six months of the year.

A timeless waterhole (BELOW) *and a totemic pandanus silhouetted against a setting sun* (RIGHT) *symbolise the quiet calm of Kakadu National Park.*

A network of billabongs along the floodplain provide refuge, food and shelter for many birds and animals. Fallen leaves and spreading canopies give some the air of English country gardens (OPPOSITE). *The trailing aerial roots of a banyan tree* (ABOVE) *anchor the branches to the earth and a watercourse* (BELOW) *reflects its overhanging canopy.*

Like huge building blocks, slabs of eroded sandstone (ABOVE) provide shelter on the edge of the escarpment.
The curious magnetic anthills of the Northern Territory (BELOW) line themselves up from north to south,
exposing their flanks to the full rays of the sun. The flower of a cockatoo apple (RIGHT) glows in the
late afternoon light.

Sharing a colour in common are fringe myrtle (ABOVE LEFT), scented wild gardenia (ABOVE RIGHT), a cockatoo apple (BELOW LEFT) and a water snowflake (BELOW RIGHT) which amazingly flowers every day during its flower cycle.

OPPOSITE: The love in the mist passionfruit is a naturalised species from South America.

A tree's new leaves sparkle (ABOVE). Bossiaea (BELOW LEFT) *and seed capsules*
(BELOW RIGHT) *add their pastel hues to the scenery.*
OPPOSITE: Hibiscus *seed capsules* (ABOVE LEFT) *and a flame tree flower* (ABOVE
RIGHT) *are highlighted by a lotus lily* (BELOW).

A lotus lily (ABOVE AND BELOW RIGHT) *contrasts with a Giant lily* (OPPOSITE) *and the flowers of a freshwater mangrove tree* (BELOW LEFT).

Naturally, a habitat as richly diverse as this sustains a broad spectrum of birds, mammals, reptiles and amphibians. On the escarpment, peregrine falcons soar on the wind draughts funnelled skywards by the ravines. The black wallaroo and its close cousin, the euro, graze in the stunted woodlands. The carpet, oenpelli and children's pythons embrace trees and rocky ledges as they patiently wait for prey to ambush. A kaleidoscopic variety of geckos and lizards are busy gulping up insects and tiny amphibians. The woodlands are favoured feeding areas for the banded pigeon, the helmeted friarbird, the rainbow bee-eater and the dusky honeyeater. The freshwater crocodile, *Crocodylus johnstoni*, suns itself along the river banks. This tropical Eden is repeated with amazing variations through each of the main vegetation areas.

The diversity is superbly demonstrated among the birdlife. The area harbours an estimated one-third of Australia's bird species, including the endemic banded fruit dove, the chestnut-quilled rock pigeon, the white-throated grasswren and the white-lined honeyeater. Parts of the park, such as the paperbark forests when the melaleuca trees flower, are essential in the food cycle of nomadic species of honeyeaters, lorikeets and other nectar-feeders and insect-eaters. The sandstone monsoon woodland is critical to some species because of its limited extent and vulnerability to damage by buffalo and pigs.

The coastal section of the Alligator Rivers region is perhaps Australia's largest virgin wetland and is frequented by massive numbers of waterbirds. It is the major refuge of many tropical waterbirds such as the pied heron, the pied goose, the wandering whistling duck and the green pygmy goose. Large flocks of juvenile pelicans add to the cacophony of duck calls. The swamps and billabongs of the lower reaches of the rivers are also refuges for incredible numbers of wildfowl. These mammoth concentrations are found nowhere else in Australia.

A lilytrotter (LEFT) *surveys its watery domain, and an Australian pelican* (BELOW) *sallies forth leaving a gentle wake.*
OPPOSITE: *A Javan file snake* (TOP) *in typical languid repose. Its pose might be mimicked by the saltwater crocodiles* (CENTRE AND BOTTOM) *but it belies their vicious nature.*

A howling jackass (TOP LEFT), *a member of the kookaburra family, waits patiently. Yellow-billed spoonbills* (TOP RIGHT) *and pied geese* (ABOVE) *take flight over a billabong.*

OPPOSITE: *Water buffalo* (TOP) *and wild cattle* (CENTRE LEFT) *have created havoc in the Kakadu National Park with their heavy-hoofed grazing. A black falcon* (CENTRE RIGHT) *glares balefully and a darter* (BELOW) *spreads its wings to dry.*

A black falcon (ABOVE LEFT) soars majestically in search of prey as a regal sea eagle (ABOVE RIGHT) takes flight. A pair of brolgas (BELOW) parade through a sea of morning mist as the sun rises above the floodplain (RIGHT).

This land of heat, humidity, glaring sun and vivid colours undergoes startling changes when the wet arrives each December. The rivers flood, the escarpment is a sheet of waterfalls and the frogs croak in day-and-night celebrations. The greening of the plains is rapid and many insects and flowers complete their life cycle during the three month wet. Not everyone, however, celebrates the arrival of the rains – the dusky rat *(Rattus sordidus colletti)* flees to the escarpment each year to escape the floods.

Other creatures welcome this short season and thrive during it. The monsoon encourages the hatching of caddis flies, mayflies and dragonflies, as well as certain grasshoppers, moths and butterflies. They all belong to the extremely diverse invertebrate fauna of the area. There are an estimated four thousand new species among the ten thousand or more species of insects found here.

This richness of live food promotes large populations of freshwater fish and frogs. And nature generously provides insect species small enough to be gulped down whole by the tiny frog, *Litoria dorsalis,* which when fully grown is only 9mm long. It has a slender, elongated body and although it has no webbing between its finger and only a hint between its toes, it favours swampy areas. It is, however, a phenomenal jumper and is often taken to be a grasshopper when disturbed.

The freshwater billabongs harbour considerable varieties of fishes which are no less remarkable zoologically than some of our mammalians oddities such as the platypus and the marsupial kangaroo rat. Among these is the saratoga, a fish of ancient lineage that shares with the Queensland lungfish the distinction of requiring a land connection at least 150 million years ago to explain its presence in Australia. All other Australian freshwater fishes have evolved from marine forms and are much more recent in origin.

The morning mist (BELOW) *silhouettes the pandanus fringing the escarpment.*
OPPOSITE: *Monsoonal rains flood the plains* (ABOVE) *and the billabongs* (BELOW).

The rise and fall of the seasonal flood levels dictates the life of many migrant birds. The lakes and billabongs often survive from wet to wet.

Many areas in the park are of special significance to the Aborigines. According to Aboriginal lore, during the 'Dreamtime' the spirit beings travelled over the landscape, giving it shape and feature where previously there had been none. These areas are imbued with spiritual associations and are considered sacred sites. In Kakadu many areas are connected with the Rainbow Serpent and are considered dangerous, and are prohibited to most tribesmen. At Djidbidjidbi and Dadbe, in the vicinity of Mount Brockman, a series of alternating dark red–black and whitish vertical water flow marks are said to represent the blood of the rainbow snake and a permanently filled rock hole on top of the cliff is considered to be the permanent home of the snake.

One of the snake sites is at Kudjamarndi in the Myra Falls area. Here the simple myth tells the story of this once popular camp site where many different tribes used to meet for ceremonies. It tells of an orphan who was camped there with a group of people and cried for his mother. His relatives offered him different types of food but he still kept crying for his mother. They told him to stop crying or the Rainbow Serpent would come and get him. He kept crying and the Rainbow Serpent came and devoured all the people.

Late afternoon. We shelter under a large rock overhang as the first of the monsoon rains fall. It is a steady, heavy tattoo. The dry soil of the lowlands beneath us is soon a river of red mud. The walls of the gorge take on a golden sheen. The trees glisten as the rain washes the dust off their dehydrated leaves. Behind us the rock wall is festooned with a marvellous display of brown and white ochre paintings of barramundi, saltwater crocodiles, a turtle and a large python. They begin to take on a rich hue in the crisp light of the sinking sun struggling to be seen through the heavy mist of rain. The wet has begun and its cleansing waters will flush and revitalise the huge wilderness that is Arnhem Land in an annual ritual that celebrates the orderly passage of time.

Rock paintings abound in almost every nook and cranny of Kakadu National Park. A man and a platypus (LEFT), an angler, barramundi and a turtle (ABOVE RIGHT) are found near Mount Brockman (BELOW AND BELOW RIGHT), a sacred site.

The pink devastation of fire (ABOVE) is reflected in the harmless setting of a sunset (RIGHT) in this wilderness in the untamed north.

Roes Rock (ABOVE) *rises starkly from an ocean of forest.* FOLLOWING PAGES: *The western Barrens dominate the horizon at sunset.*

THE SALT DESERT

Fitzgerald River National Park, Western Australia

LOCATION: *Approx 300 km east of Albany.*
SIZE: *250 000 hectares.*
GEOLOGY: *Dramatic changes from north to south with granite, then spongolite and finally quartzites predominating.*
CLIMATE: *Mild coastal climate with winter rain season.*
FLORA: *Extremely diverse flowering shrubs. Heathland predominates.*
FAUNA: *Rich birdlife but few mammals.*

FITZGERALD RIVER NATIONAL PARK, WESTERN AUSTRALIA

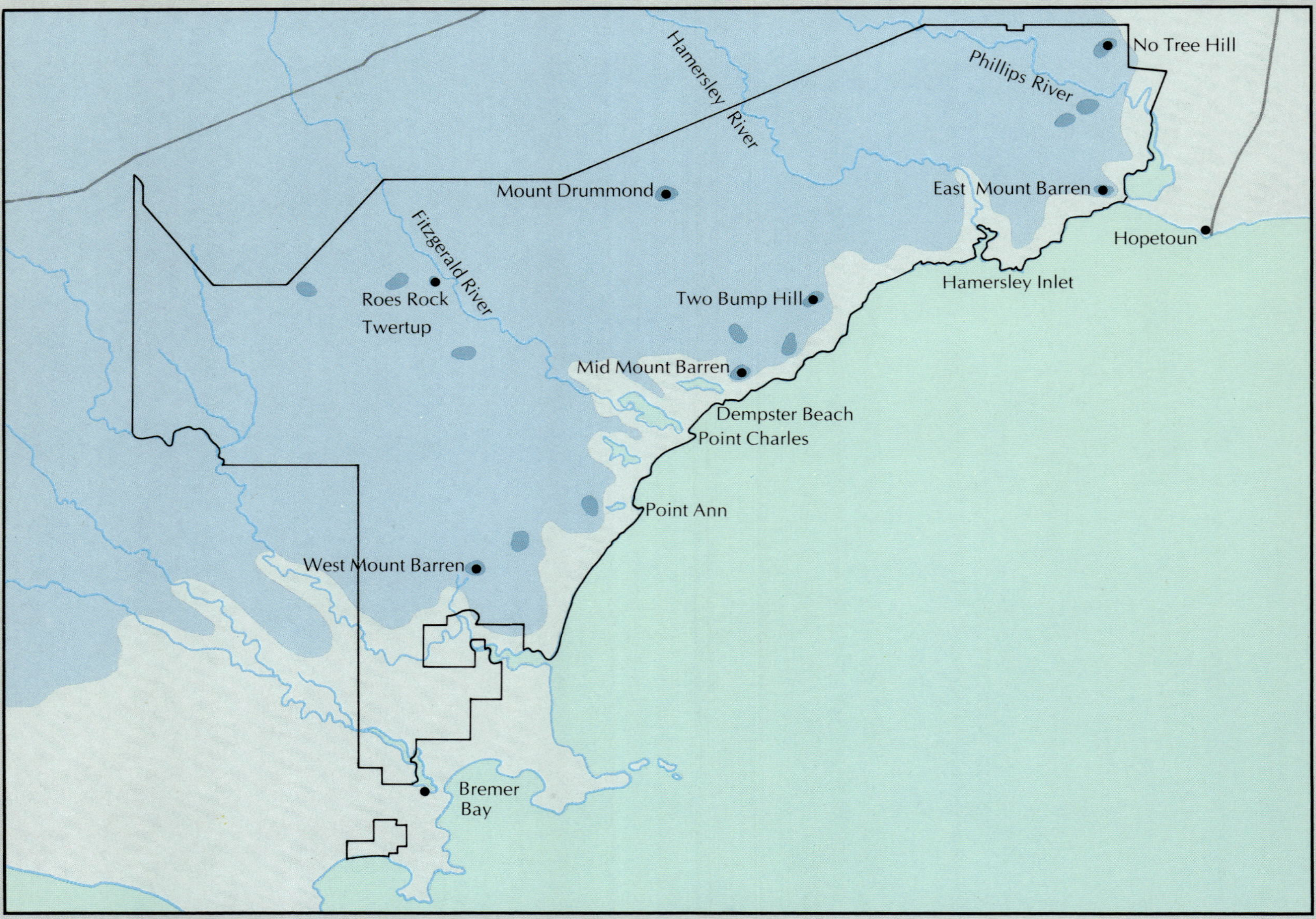

Hamersley River

Phillips River

No Tree Hill

Mount Drummond

East Mount Barren

Hopetoun

Fitzgerald River

Roes Rock
Twertup

Two Bump Hill

Hamersley Inlet

Mid Mount Barren

Dempster Beach
Point Charles

Point Ann

West Mount Barren

Bremer Bay

I N A QUIET VALLEY cradling the middle reaches of the Fitzgerald River there is a salt-fringed sepia stain cascading down a huge granite face – a waterfall frozen in time. Past floodwaters, swollen with red mud and the briny legacy of the saltlands, have left their mark. The waterfall links two large pools scalloped by erosion in the red granite base of the riverbed. Their banks are fringed with green lichen – paper-thin lawn rooted in rock. Stringy plants vie for survival in pockets of windblown soil and debris trapped in clefts and cracks.

This harsh environment, typical of the Fitzgerald River National Park, clings to the belly of Western Australia. Its 250 000 hectares encompass a bleak landscape of spongolite cliffs honeycombed by the elements; scraggly dry streambeds outlined with salt tidemarks; bronze-toned granite flecked with red flashes of weathered flaking rock; a coastal barrier of sand dunes as white as snow; broad valleys carpeted with low scrub, paperbark, tough mallee and wiry melaleucas; and three low ranges each topped with a small abrupt summit named Barren. The three peaks – West, Mid and East Mount Barren – are remnant islands, once barely 90 metres above sea level.

Their existence is responsible for the botanical abundance found in the threatened boundaries of the National Park. While still islands in an ancient sea, they provided refuge for the fertile flotsam of the plant world seeking anchor during the mammoth upheavals of the Proterozoic period some 3000 million years ago. This was the age of Gondwanaland, the super continent formed when the world's land masses were one and the fertile Fitzgerald area was joined to Antarctica.

PRECEDING PAGES: *Weather-sculpted granite along the upper reaches of the Fitzgerald River. Wind-scoured spongolite cliffs* (BELOW) *glow as they reflect the late-afternoon sun.*

*Nature's forces — wind, rain and unrelenting sun — have created
a bizarre combination of eroded granite shapes along river valleys
in the northern section of the park. A pitted rock (TOP) marks the
gravesite of an ancient Aboriginal burial site. The dramatic temperature
contrasts of the area have created a network of expansion cracks (LEFT)
which often provide refuge for wildlife, like the green-spotted frog
(ABOVE). Salts carried from the central floodplains have left white
tidemarks (RIGHT) as trapped water evaporates.*

Mid Mount Barren (ABOVE) is the highest peak of the central mountain range that bisects the park. From its summit ridge (RIGHT) the curious effects of millenia of erosion can be seen traced on nearby hills. Spongolite outcrops (BELOW), their summits once part of the seabed, dominate large portions of the central part of the park.

Today, some seven thousand plants species are found in the vast expanses of Western Australia and the Fitzgerald River National Park boasts more than thirteen hundred of which at least seventy are endemic. The area is considered the cradle of Australian plant life and the unique flora have been protected and isolated by the equally formidable barriers of vast oceans and the bulk of the driest continent.

The wide valley of the Fitzgerald River (a meandering, occasionally connected string of brackish pools) is a palette of muted colours – the legacy of the cornucopia of plant riches: it hosts every shade of green from lime to emerald, with a host of greys, reds and blues reflecting off individual leaves. Straw- and beige-coloured dry leaves, often edged with purple, add to the canvas.

The flora share a major characteristic with the fauna found in the park for the plants, like the important endemic mammals, birds, reptiles and insects, are only found in specific areas. The variable mosaic nature of the geology, soils and vegetation has produced a complex pattern of distinctive and isolated habitats. The flora and fauna have evolved in this mosaic environment, resulting in small, discreet populations in small, discreet locations.

The northern fringe of the park is basically granite – old granite, weathered, worn and battered by the elements for an estimated 1700 million years. Today it wears its aged patina of sooty grey with the nonchalance of a dowager. Time has scattered mammoth blocks of stone down the river valley and left them in delicate balance. Pitted footprints of time etched in their surface trap scarce rains. Skeletal driftwood bleached by sun sparkles in contrast. This is where our journey begins.

A spiny echidna forages among stunted trees on the exposed slopes of Mid Mount Barren. Its spiky armour is mimicked by the unique resurrection plant (ABOVE RIGHT). This plant demonstrates remarkable survival powers in this harsh environment and its colour is often an indicator of the moisture of the soil. With the slightest of rainfalls, it turns green. In an area like this with its spectacular variety of plant species, even the seeds compete for attention, among them (BELOW RIGHT) seeds of the aptly named Cyclops melaleuca.

A large tiger snake, belly bulging with a recent feed, lies sprawled in the sun, its body a scaly and grotesque imitation of a mallee root. Disturbed, it sluggishly eases into a recessed overhang. When it emerges again it takes on a more elegant curve – a coil under tension.

A hole, drilled into the riverbed by the grinding, crushing action of rocks tumbling and roiling during flood, sinks two metres into solid granite, its sides sheer and smooth. The pool of water it holds at first appears to be the greasy black of the bottomless pits found in some canyons until a break in the cloud cover reveals 50 centimetres of clear water covering a nest of rounded rocks of all colours. A turtle's head riffles the glassy surface. It takes a hurried breath and sinks down past the stiff corpse of a speckled goanna which lies motionless, limbs languidly outstretched – a peaceful scene almost mimicking a Victorian glass bowl enclosing stuffed animals pinned in stilted poses. The goanna, a recent victim of its own over-curious mind, will be fodder for the turtle. The survival of the turtle is a matter for conjecture: will it be able to withstand the churning hell that flood-waters produce in its isolated haven?

Nearby, Christmas spiders in gregarious chaos string their webs across windswept gaps close to the ground. Their spiky bodies, neatly patterned with triangles of orange, yellow, black and white, are visual oases – a striking effect recreated at dusk as a freak shaft of light embraces nearby Roes Rock.

A steep-sided hole (BELOW), *scoured out of granite by the tumbling boulders brought downriver by infrequent floods, becomes a fatal trap for the unwary or merely curious. A tiger snake digests a recent meal whilst basking in the sun. A cornered goanna* (BELOW RIGHT) *demonstrates its threat display by posturing stiff-legged and hissing.*
OPPOSITE: *A spongolite cliff shows the ravages of time and weather.*

Low-slung clouds cause divergent rays of sunlight to diffract through the dust-laden air like a heavenly pronouncement. The flat top of Roes Rock anchors the panoramic sweep taking in the Fitzgerald River valley and the coast. Pregnant clouds, heavy and rain-threatening, slink inland to tease the farmers on the park's northern borders. Slivers of sunlight race across treetops like breathing surf. A cool sea breeze rustles the banksia seed cones in a murmuring chorus backed by the gentle warble and song of the birds preparing to settle for the night. It is a treasured wilderness time-warp – a serene end to the day. Then suddenly it happens, a grand and unexpected finale! The sun, for a minute, pierces the broody cloud canopy and its near-horizontal sweep picks up the pale spongolite cliffs of Roes Rock. Like a diamond it shimmers in the spotlight – a lone sparkle in the grey-green gloom of dusk. Too soon, it disappears.

The geology and ecology of the park have been studied by George Kendrick, assistant curator of the Department of Palaeontology at the Western Australian Museum. He found that the Pallinup siltstone (or spongolite) is mostly made up of deeply weathered sponge remains and that the sponge bodies have disintegrated along with the weathering. As a result, the sponge bodies are like Lewis Carroll's Cheshire cat – they can not be seen but their 'grin' remains! He also found fossils of bivalve shells, possible internal moulds of sea urchins, a marine worm burrow and remains of terrestrial plant leaves. The evidence suggests that rivers were actively discharging forest litter into the sea. Rainforests probably covered most of southern Australia in the late Ecocene period, some 65 million years ago.

Dusk mutes the harsh outlines of the Barren Ranges (BELOW).
OPPOSITE: *Lichen-encrusted granite* (BELOW RIGHT) *shown in the same light has the sombre quality of a graveyard, while Roes Rock* (TOP AND BOTTOM LEFT) *sparkles in the last light of the day.*

Nature's harsh cycles of life and death never take a holiday.
This march fly (LEFT) ended its short life in a spider's web; it could
as easily have been trapped in the deadly claws of a praying mantis
(BELOW). The harshness of life in this environment is offset by the
sparkle of the plants – hibiscus seed pods (ABOVE) and the
adenanthos flower (RIGHT).

A wealth of plant life abounds in the Fitzgerald River National Park, which has justly become famous for its abundant and vibrant flower variety. Typical are the beaufortia (ABOVE LEFT), the red toothbrush (Grevillea) (ABOVE RIGHT), the Melaleuca elliptico (BELOW) and the pea bush flower (RIGHT).

Diversity in shape, size and colour has created a lush botanical bonanza: the silky-leaved blood flower (ABOVE) *and a eucalyptus in flower* (BELOW).

OPPOSITE: *(Top to bottom, left to right) Lambertia, an as yet un-named species, purple beaufortia, dryandra, new leaves of a native shrub, a peabush, a eucalyptus seed pod, the sturdy sea rocket and another dryandra.*

Spongolite looks and feels like aerated sandstone – an ideal modelling agent for the elements. Wind, rain, heat and cold in turn file, scrape, erode, crack and blister the cream rock into bizarre shapes – some smooth and rounded with the pregnant bulk of Henry Moore sculptures, others bristly like stone echidna or jagged like the serrated jaws of a white pointer. A lump the size of a concrete building block barely weighs as much as a house brick.

The Twertup Field Study Centre, near the geographical centre of the park, is situated on a spongolite cliff once 20 metres under the sea. In aeons past, sea levels ranged from 140 metres higher to 80 metres lower than present levels owing to changes in the volume of the earth's seas. Spongolite is a sedimentary rock and at Twertup it is estimated to have been formed about 40 million years ago. The spongolite is, however, young in comparison with the quartzite spread through the Barrens, south and south-east of Twertup. Here the rock is an estimated 1400 million years old and the granite found alongside the pool near Roes Rock has been dated at 1700 million years. (The first known white person to pass through the area was Surveyor-General John Septimus Roe in 1848.)

Over a few thousand years, humans have developed a tremendous power through the explosion of their culture. Although in the geological time scale these years are a mere hiccough, humans actually have become a geological force, like glaciation or vulcanism. Not only are they capable of altering the landscape and the balance of the ecosystem – they *have* altered it. The Fitzgerald River National Park is surrounded by human landscaping. Under virgin conditions, the sub-soils of southern Western Australia contain significant quantities of soluble salts. Since agricultural development razed the forests, scrub and heath, hydrological changes have caused the salts to concentrate in the surface soil in low-lying parts of the landscape. As a result, about 250 000 hectares of cleared agricultural land have become unproductive. 'The salt problem has proved over the years to be intractable,' says C. V. Malcolm, a senior research officer with the Western Australian Department of Agriculture.

The powerful sculpting effect of the weather on the relatively soft surface of a spongolite cliff (LEFT, BELOW AND TOP RIGHT) *creates a serrated ocean in stone. Trapped in a pool of brine, a branch* (BOTTOM RIGHT) *soon becomes a salt-encrusted effigy.*

A hardy, leathery-leafed tree (LEFT) has somehow gained a foothold on a quartz cliff face, and while certainly not thriving, it nevertheless survives with tenacious vigour. Similar stamina is displayed in the harsh beach environment where salt-laden spray and sand encrust a fleshy pigface, (ABOVE) and hardy arctotheca (RIGHT).

Yet the problem of salt concentration after removal of trees has been known since early this century. In 1907 the *Journal of Agriculture* published an article in which it was reported that it had been 'pretty conclusively proved' the removal of trees affected water supplies since rainfall passing through the soil took salts with it. In order to prevent salting, it considered it would be necessary to replant a high proportion of the trees that had been removed. In 1917 the same journal published another article which claimed that forest scrub influenced salt behaviour by using water and thereby stopping salt rising. Yet at the time of writing, the Western Australian Government was planning to release several thousand hectares of virgin land on the park's northern border for clearing.

In his book, *The Forest and the Sea,* biologist Marston Bates notes: 'Man has not escaped from the biosphere. He has got into a new, unprecedented kind of relationship with the biosphere: and his success in maintaining this may well depend not only on his understanding of this world in which he lives ... it looks as though, as a part of nature, we have become a disease of nature – perhaps a fatal disease.' In a classic understatement, he adds: 'I am not advocating a return to the neolithic... but long run efficiency would seem to require certain compromises with nature.'

During the night the Milky Way dissects the moonless sky's black void with a scatter of brilliant white gems. A clutter of tin cans hurtles in computer-controlled orbit across the sky to the north, like a nightmare producing vivid images of the beer cans that litter our highways today being transposed to some future skyway – permanent satellites of the slobs of tomorrow.

Morning. A rim of red ignites the horizon to announce another wilderness dawn. It is only the curtain-raiser, the sun itself is another twenty minutes behind. The ground is damp from a light dew. A metre away, a spongolite cliff drops off into a flat valley.

The bad dream was still there as we went for an early morning stroll. We felt a special awareness of the fragility of the world's ecosystem. Parks like this are under so many threats – from farmers demanding more land to weekend cowboys, strapped into bucket seats atop three hundred horsepower steeds, looking for bigger sandpits to destroy. In this park, streams and creeks are being severely polluted by fertiliser run-off from mono-cultured fields to the north and the salt-laden water table is being raised by crops unable to use as much water as the scrub and heath they replaced – a real threat to the shaved fields and an ecological disaster for the indigenous flora.

We walk along the cliff edge and stumble on a tree root. It belongs to a crippled moort tree *(Eucalyptus platypus),* its base a shattered explosion of weathered wood shards, flaked bark and twisted roots. Falling victim to wind and the erosion of the shallow layer of topsoil capping the spongolite base, it had virtually been ringbarked by its fall but the narrow strip remaining had managed to drip-feed the shattered remains. The flow of life fluids was obviously insufficient to save many branches, now reduced to stark grey fire fodder; their sacrifice, though, has ensured the tree's survival. Today, close to the original lower branches, a new tree has emerged – vibrant with vigorous growth, even though its future depends on the twisted and mutilated root system and tenuous sap feed from the damaged bark layers of the original tree.

We continue our walk. Two owls show swivel-necked surprise at our intrusion before flying off ten metres. A shy, brush bronze-wing pigeon preens itself on a sunny shelf in an obvious ritual. The river calls…

The Fitzgerald's sporadic downhill journey is a tortuous meander over granite, spongolite and quartzite. Each offers its own challenges and resistance to the downhill flow of the river's saline solution which finally spreads itself, just before reaching the sea, in a wide and shallow basin reminiscent of the saltpans of central Australia. The basin's shimmering heat hazes create confused mirages offering false succour. Reality is a salt-laden, dried-up oval littered with pieces of sun-bleached driftwood squatting rigidly, stark branches reaching up in supplication.

A river of salt (LEFT AND BELOW) *is the legacy of the rising watertable of the southern section of Western Australia. The clearing of millions of hectares of natural shrub for wheat farming has seen the salt-laden watertable lift and infect the streams and rivers with its briny content.*

Nearer the mouth of the river lies a large, placid lake framed by white dunes to the west and coarse scrub on the eastern bank. Its edges are quicksand in solution over a bed of black rotting vegetation. The quagmire is a foretaste of the oozy brine in the lake itself – a supersalinated solution, far richer in salts than the ocean itself and so dense you can float without effort.

To reach the sea, the Fitzgerald has to breach a low-lying, half-kilometre-wide bar of white sand. 'Prevailing winds have pushed the beach sands into high sand dunes – blisteringly bright hills reshaped continually by the incessant wind. Further east, these coastal dunes challenge the pinkish dunes of eroded spongolite marching south. The sands do not mix: the dividing line is clear and crisp, but writhes with the convoluted agony of eternal battle.

The wind attacks both sets of dunes with unbiased vigour and sends corrosive sprays of fine sand blasting off the crests like aerosol sprays. At lower levels the wind coats tenacious grasses and shrubs with a thick layer of salts and sand: in places it slices wet sand like a knife leaving a pockmarked Swiss cheese effect. Elsewhere, the constant whip of branches and grass stems have created half-moon arcs in the sand, leaving large areas looking like the scaly carcass of a mammoth deep sea denizen. Despite these forces, small fronds of delicate lace seaweeds – their pink and cream and lime hues glowing – dot the gap between high and low tides in an apparent no-man's-land for the wind. The tiny float nodules of other species of seaweeds punctuate the limits of the waves drying on shore. Like the messages in Morse code, the nodules stretch from one end of the narrow belt of beach to the other.

Wind lifts a flurry of sand off the top of a coastal dune (BELOW) *as it persists with its daily reshaping of millions of tonnes of sand. Spray-laden sand coats some hardy grasses* (ABOVE RIGHT) *that have survived the stinging lash of sand lifted by the constant winds* (RIGHT) *that batter this exposed coastline.*

Behind the first rank of coastal dunes, a strip of melaleuca forest thrives, spreading a wide canopy of shade – a green roof of robust clusters of leaves neatly meshing together in a precise mosaic. It is a tranquil setting that hides the harsh reality of survival in the bleak terrain. Below the canopies, the dense shade has killed all competing foliage, including juvenile branches of trees that have won the battle against neighbouring trees for the sun. The battle, aided, abetted and some-times thwarted by storms, fires and natural decay, creates a spider's web of branches as trees send spreaders out to take advantage of a neighbour's disaster. There are no winners but the species survives.

Dawn. The Southern Ocean is a pink pond barely breathing with the rhythm of the light swell. As the summer sun clambers higher, it undergoes a subtle trans-formation through the full range of pastels. Finally it settles on a cerulean hue which deepens into aquamarine. The shallows – a floor of pure white sand – filter the deep blue as the sun probes. Silvery slithers of salmon drive through the swells. Closer in shore, occasional breakers lift the fish into silhouette. Wading out, clouds of sand rise from under-water footprints. Our journey down the Fitzgerald River has ended.

We visit a hidden valley in the Hamersley River gorge – an oasis of green with still ponds and prolific bird life hammering out a constant chorus. Avian exotica abounds... raucous white-tailed black cockatoos, busy bee eaters, wood swal-lows, speckly honeyeaters... the white-naped honeyeater, the red-eared firetail and the spotted pardalote have been sighted nearby. A fantail visits. Scruffy, cheeky, he dive-bombs the human intruders. Feathers awry he alights on a nearby branch and with tilted head sticky-beaks. Sunset brings grand reflections and there is fire in the river pools reflecting the highlighted treetops. The new moon brings added pleasure. In the background is the soft roar of surf.

The coastal melaleucas engage in a bitter battle for sunlight (BELOW) *that creates a jig-saw effect in the canopy.*
OPPOSITE: *A square-tailed kite* (ABOVE) *soars with swivel-necked ease as it scours the terrain for food. Meanwhile, a small frog seeks cool relief in a pond of fresh rainwater. A tranquil setting* (BELOW) *disguises the salt legacy of the park's waterways.*

The royal hakea (ABOVE) *thrives in the eastern section of the park,
as do the wispy blue weeping gums* (BELOW RIGHT). *South of the park
lies the broad sweep of the Southern Ocean* (RIGHT).

Later, at the the mouth of Dempster Inlet, we are besieged by persistant silent mosquitoes. The sturdy flesh-stabbers ignore the rich chemical brew of modern insect sprays and attack.

We travel on and discover a tributary of the Fitzgerald. Trapped in shallow pools above a layer of impervious clay, its sprawling edges are a tidal crust of thick, pure salt. The water, laden with salt, is like glass. It embalms detritus and inquisitive insects. On a flat shelf, wind-driven spray has woven a tapestry of red, purple, brown and white from the dried-out salts.

At the eastern edge of the National Park we find No Tree Hill, a rocky lump pimpled with the royal hakea and bearded wispy blue weeping gums. Known as *Eucalyptus sepulcralis* (from the Latin word for grave), the weeping gum was discovered and labelled in 1882 by Ferdinand von Mueller, who wrote that the name was chosen 'because this eucalypt will be destined to add another emblem of sadness to the tree vegetation of cemeteries in climes similar to ours.'

The botanical abundance of No Tree Hill typifies the mosaic structure of the unique fauna and flora for here conditions permitted the survival of plant life so immeasurably old. Primitive forms have survived untouched through aeons of time or have evolved quite differently from plant life in other, less tranquil, areas.

The royal hakea *(Hakea victoriae)* is only found in the Barren Ranges and is a fine example of the endemic species. A hard-leafed plant, it feeds on itself: each year the new growth takes food from the layers of leaves below it, and the varying degree of its incestuous appetite gives rise to a graduated colour display. Fresh growth is a combination of green and yellow: last year's layer is red and orange, the one below is orange and brown, and those that stay further down are the uniform grey-brown of dead leaves.

A thunderstorm heralds our return to the valley of the stone waterfall. It is a farewell visit. The sky is an ominous black boiling mass, grumbling with thunder. We shelter under a rock ledge and enjoy the show.

The crack of lightning and the slow roll of thunder dominate until the setting sun suddenly breaks through under the storm and pinkens its swollen belly. Suddenly the storm is vulnerable and, as if it realises this, it soon peters out. We venture into the rock-strewn gully. A wet sheen has awakened the colours of the lichen and rock. The waterfall, frozen in time, shines and shimmers but still does not flow.

The southern border of the Fitzgerald River National Park sees the meeting of a harsh untamed landscape and an often brutal ocean. It is rugged terrain, but it still supports a healthy variety of plants — stubborn, hardy plants that exist despite constant abuse by gale-force winds and searing heat.

Rare rains deepen the rich reds, ochres and browns of the Simpson Desert.

THE LIVING DESERT

Simpson Desert, South Australia

Sulphurous deposits (ABOVE AND RIGHT) *mixed with mud and sand provide a palette of muted colours on the fringes of a hot spring.*
FOLLOWING PAGES: *Pale, salt-encrusted claypans isolate islands of red sand.*

The timeless grandeur of the desert is particularly evident from the air and here (ABOVE AND RIGHT) *the effect is enhanced by freshly formed shallow lakes.*

LOCATION: *Northern South Australia.*
SIZE: *150 000 square kilometres. spreading through South Australia, the Northern Territory and south-west Queensland.*
GEOLOGY: *Mainly parallel red sand dunes running north-north-west to south-south-west on vast sedimentary basin.*
CLIMATE: *Dry and hot with less than 250 mm rainfall most years.*
FLORA: *Mainly small, hardy shrubs and perennials superbly adapted to the harsh environment (see text).*
FAUNA: *Surprisingly diverse collection of reptiles, insects and birds with many tricks for survival (see text).*

SIMPSON DESERT, SOUTH AUSTRALIA

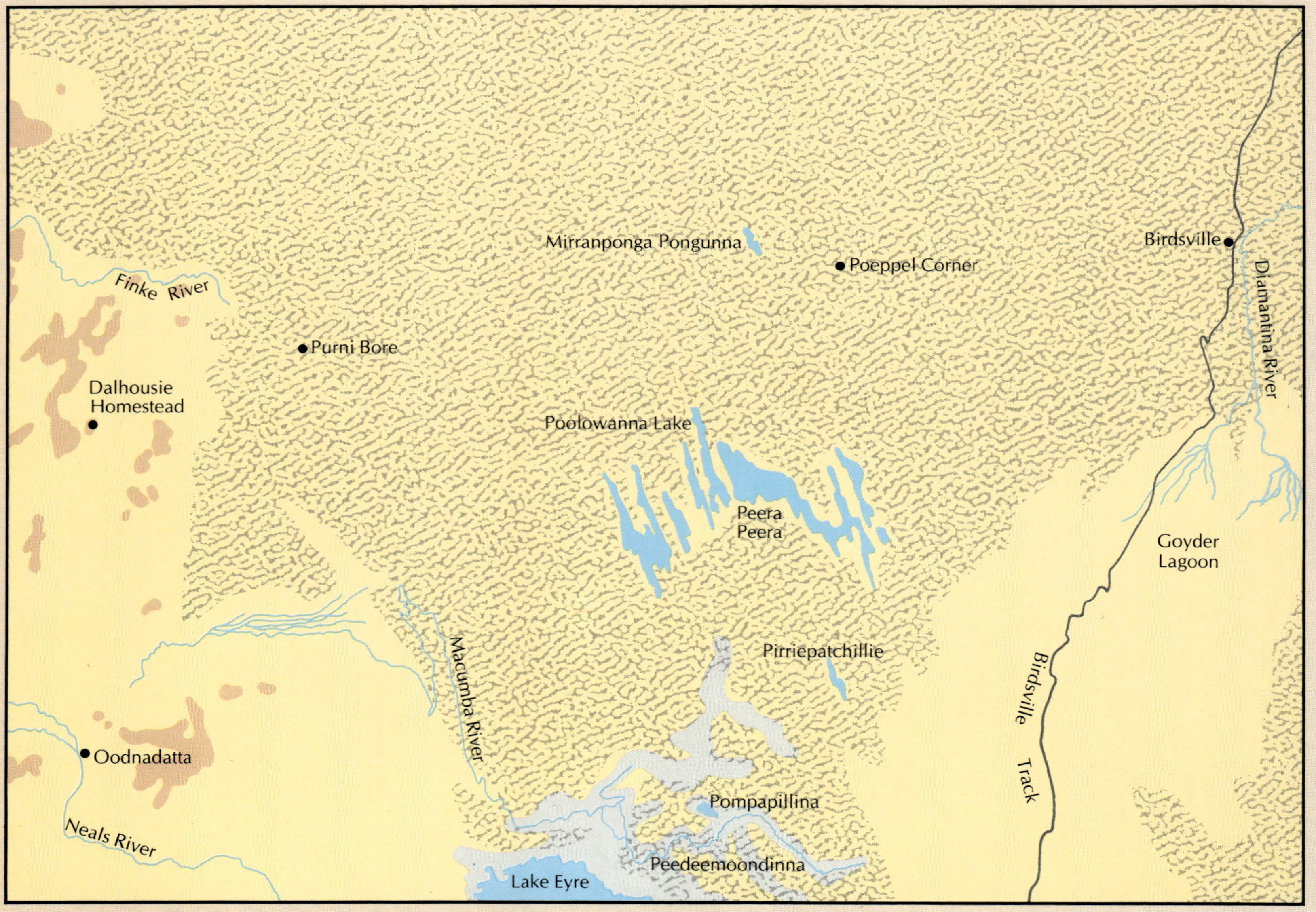

Finke River

Purni Bore

Mirranponga Pongunna

Poeppel Corner

Birdsville

Diamantina River

Dalhousie Homestead

Poolowanna Lake

Peera Peera

Goyder Lagoon

Macumba River

Pirriepatchillie

Birdsville Track

Oodnadatta

Pompapillina

Neals River

Peedeemoondinna

Lake Eyre

THE WHIRLWIND SPIRALS across the crest of the red sand dune and vacuums the loose sand upward into the heat of the midday sun. To the north and south mirages beckon with their false promises of shimmering water. A solitary brown falcon, disturbed, takes to the air and leaves its shadow to crisscross our path. The baked earth between the dunes is like the aftermath of battle. Tiny skeletons pebble the valley, and burrows not unlike bomb craters pockmark the bare patches of soil between scattered mulga and spinifex. Not a track, a sign of movement, is visible. It is quiet, the deathly hush of the desert by day.

From the top of the dune the Simpson Desert spreads in all directions, an unrelenting sea of sand waves that will never break on an ocean beach. Stabilised by wiry desert plants, the long parallel dunes with their red crests strut across an ancient seabed drained through cataclysmic upheavals 60 million years ago. The sands of the Simpson Desert are the wind-deposited debris of the mountain chains that pierced the once-fertile plains of Western Australia during the Jurassic Period, when Australia was still part of the supercontinent Gondwanaland. Decimated by the erosive action of the elements over many millions of years, the red mountain dust now mocks the former Inland Sea by imitating its wave action.

The sun climbs to its zenith, the thermometer registers 45° Centigrade in the shade. More whirlwinds form. Most last a minute or two before succumbing to lethargy; others appear to gain perverse strength from the barren terrain and suck up tubes of sand which then go waltzing up and down dunes. There is an emptiness around us, the desert is lifeless, the sun a burning orb in a cloudless sky.

The rich red soils of the desert are constantly on the move as persistent westerly winds sculpt the fine sand into waves.

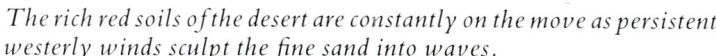

The heavy tread of cattle (ABOVE) has churned many muddy waterholes into a cement-hard surface of pimply clay crusts, leaving only salt-laden pans (BELOW) undisturbed. The harsh reality of the Simpson Desert is reflected in the barren shrub (ABOVE RIGHT) and in the mud-caked skeleton (BELOW RIGHT).

An endless sea of sand (ABOVE) stretches from horizon to horizon, red crests separating exhausted patches of mulga trees (BELOW). Tough, wiry grasses with intensive root systems (RIGHT) strive for survival in this forlorn ocean of sand.

Shadows slowly lengthen, the eastern sky changes colour in its time-honoured transition from day to night. Twilight does not linger and soon the desert sky sparkles with galaxies of stars. The air is still. All is quiet. It is at least five degrees cooler in the valley than on the sand ridges which slowly radiate the heat they have absorbed all day.

A light breeze heralds dawn and a sky full of thick clouds. The sand ridge is a filigree of delicate tracks – the comma-like skids of a snake, the evenly spaced hops of a small kangaroo rat, busy, erratic signatures of lizards, geckos and goannas, tiny lines dotted by the many feet of centipedes, and the lacework of an evening's labour by ants, beetles and scorpions. It has been a busy night in the living desert.

The fauna and flora of the Simpson Desert have evolved in bizarre ways to survive the harsh conditions and their numbers are a celebration of life. There are frogs that bury themselves with an internal supply of water to tide them over droughts (enterprising Aboriginal nomads soon become acquainted with them and they would dig them out and squeeze them to slake a thirst!); dried-up lakes and riverbeds churn with freshly hatched fish after rain – the eggs having survived as hard cysts for years; tiny marsupials store food reserves in fat tails, and most creatures are nocturnal. On its western edge the desert also harbours the elusive hardyhead – a tiny freshwater fish of the genus *Craterocephalus* that is found only in the hot-spring oasis of the site of the abandoned Dalhousie homestead, north of Oodnadatta. Barely seven centimetres long, it is found nowhere else in the world.

Roosting birds (BELOW) *are silhouetted by a rising full moon which follows a rain-threatening sunset sky* (ABOVE RIGHT). *Morning sees a languid setting, the red sand showing evidence of a busy desert night* (BELOW RIGHT).

Rugged spinifex scrub and stunted dark-green mulga trees dominate the flora of the desert – between the rains. The symmetry of sand waves is broken by the random tracks of sliding snakes and various small animals (RIGHT).

The rains bring a bulging-eyed burrowing frog (ABOVE) to the surface. It hibernates during the dry spells, burying itself in the soil by digging a vertical shaft with its feet. A primitive crustacean (LEFT), custodian of primeval survival tactics, only emerges when heavy rains soak the desert watercourses. Cold morning air creates a steamy mist over a hot spring (BELOW) while its sulphur-laden waters create a jig-saw of fragile platforms on the surface (RIGHT).

The sulphur-laden waters of a gurgling hot spring form surreal landscapes of stone waterfalls, encrusted insects, scalloped platforms and a bubbling cone.

Adaption is the code for survival in the desert and in summer many creatures – including insects – simply go to sleep for the duration. Their metabolic rates drop so they can survive prolonged periods without food. In this dormant state they wait for cooler weather or the rare rainstorm that will bring forth fresh plant-growth.

Arid area plants are superb survivors. The humble spinifex, curse of the traveller on foot as it stabs through cloth and skin, is especially adept at survival. It has a vast root system and in many parts of the sand country it is the only factor binding the soil. In extreme drought conditions, its spiky leaves react by curling into tight cylinders, effectively cutting down its evaporative surface area – the mass of tiny stomatal pores that connect the plant's interior with the external environment. The stomata diffuse moisture to the atmosphere and are only open during daylight to allow photosynthesis to take place. (Carbon dioxide is drawn from the atmosphere and converted to tissue-building materials.) The pores must therefore be open to allow growth, but the process then allows precious water to pass from the plant. The spinifex simply accepts a stunted future and closes down half its stomata. The design of the spinifex clump also acts as a rain·trap, guiding any drops to its core where the roots are densest. Dead leaves and wind-blown debris mulch its base to protect soil moisture. Despite its harsh exterior, the tough spinifex provides welcome refuge to many species of insects, lizards, snakes, birds and small mammals sheltering from sun, wind and the omnipresent aerial predator – the brown falcon.

Other arid zone plants also owe their survival to being able to cut down on water loss. In severe conditions, the mulga tree only opens its stomata for a short period around sunrise. Its growth is several retarded, but so is water loss. The mulga's branches radiate outwards and are also an efficient rain trap. CSIRO experiments have shown that a tree five metres high and five metres in branch diameter can collect 200 litres of water from a rainfall of 25 mm. This ability also attracts a green beard of tiny grasses round the trunks of the mulga during milder conditions.

Life-saving clusters of pale spinifex (LEFT AND RIGHT) *punctuate the desolate terrain, while taller mulga offers shade and shelter also. The will to live is determinedly demonstrated by a wiry spinifex root* (BOTTOM LEFT), *while a broad mulga tree* (BOTTOM RIGHT) *shades its mulch of fallen leaves.*

The Simpson Desert presents a featureless landscape which is intimidating to the explorer. There are no landmarks to guide the traveller in this ancient world. The mountains that soared perhaps as high as the Himalayas in the convulsions of creation more than 400 million years ago are today worn stumps that barely dent the horizon. The sea of dunes going nowhere reflects the geological calm of this aged terrain. The parallel sand dunes running from north-north-west to south-south-east scour the countryside like furrows raised by a giant plough. The longitudinal ridges reflect the path of the constant prevailing winds. Barely 15 metres high on average, the dunes often stretch unbroken for up to 120 kilometres.

From the air the dunes initially maintain their monotony, but the seemingly predictable pattern of the sand ridges marching in tight rank from horizon to horizon is deceptive. The shapes that reveal themselves from 2000 metres are subtle, delicate and fascinating.

Occasionally, the prevailing winds become tired of being channelled so tightly and the corkscrewing air currents link the ridges together as they create a vortex. On the northern fringes, the dunes run square into the ancient quartzite outcrop of the Macdonnell Ranges. In the east the ridges are spaced by saltpans and claypans rimmed with precipitation rings and in the west the vast floodplain of the Finke River corrals the dunes. And then further south the red dunes gradually fade to a pale ochre as they mix with the sandy remains of the southern quartzite mountains.

Flying north from Adelaide, the vast presence of Lake Eyre soon dominates the view below. This barren saltpan broods in the blistering heat of the summer sun, reflecting wave after wave of glare from the bare white surface. It is a hostile environment that has not changed since 24-year-old Edward Eyre first sighted it in June 1840. He was leading a small expedition looking for an overland route for cattle drives to new pastures and markets. Only 600 kilometres north of Adelaide the expedition was halted by salt marshes between Lake Torrens, which he named, and another salt lake that 18 years later was given his name. The party's horses sank to their bellies in the glutinous mud and young Eyre was also confounded by the images that pranced across the lake's surface. His journal notes: 'From the extraordinary and deceptive appearances caused by mirage and refraction, it was impossible to tell what to make of sensible objects, or what to believe on the evidence of vision'.

A sea of sand dunes (BELOW) fades into the azure haze of infinity, while the moonscape of saltpans (OPPOSITE) spreads its white sprawl across the fringe dunes.

Red gibber stones are a prominent feature of the severely eroded plateaus of the western borders of the Simpson Desert. Necklaces of deep-rooted trees mark the paths of ancient watercourses.

Today, much of the land surrounding the desolate lake belongs to huge stations, where sheep have shaved the earth to its brown base. Stunted trees speckle the dry watercourses, their roots dug in deep. These convoluted pathways drain the flat and corroded surface of central Australia. There is a broad, lace-like effect etched where the creeks finally come together on the wide floodplains and where the tentacles of this mammoth drainage system finally fade, the saltpans and mud-flats wait to trap the last drops. These ancient patterns are folded in harmony with the low contours and leave a sense of balance and order which unfortunately is constantly jarred by the arbitrary fence lines that march across the landscape.

On this journey we find the Finke River in flood, its banks overflowing. The surface of the brown torrent is dimpled with pressure waves. Yet, less than a kilometre away, another riverbed lies barren – a dry dusty snake coiled on the landscape. The Finke is being fed by the aftermath of 400 mm of rain that fell on Alice Springs a few days before. The river's tortuous path winds through the distorted peaks and valleys of the Macdonnell and other smaller ranges before reaching the desert fringes. From the air these remnant mountains bear testimony to the violent forces that formed them. Molten rock and lava were spewed into layers, then tossed and kneaded into spectacular folds before they had time to cool and solidify. In the millenia since, erosive elements – mainly water and wind – have scoured away the beds of soft shale and sandstone and isolated the resistant quartzite to leave undulating ridges fanned out across the countryside like the backbones of mammoth reptiles.

There is a grotesque beauty in the timeless make-up of the face of the earth: the rouge-red cheeks of gibber, the grey-green mask of the glaucous vegetation, the claypan freckles, the rutted lines of age outlined by the single files of trees marching down the dry creekbeds and the pleated ruffles of the endless folds of sand. Sadly, the imagery is disrupted by the ruler-straight seismic tracks which crisscross vast sections of the Simpson Desert from horizon to horizon. Created in 1964 in a fruitless search for riches below the bleak surface, the scars remain twenty years later. The desert is slow to heal.

The relentless seismic tracks which slice through dunes and valleys eventually become part of the abstract pattern – a tribal tattoo on the guts of Australia. With a bit of foresight, the oil-hungry explorers could have spelled out some message for future visitors from other worlds. Instead, they have created meaningless symbols not unlike a sheet of graph paper.

The rains that fell 200 kilometres to the north several days earlier finally filter down the desert riverbeds (ABOVE AND LEFT). The north-west borders of the Simpson Desert are characterised by eroded gullies and bleak remnant plateaus (BELOW).

The lizards of the desert are another breed of survivor, superbly adapted to the drought-plagued environment and adept at defence strategy. Most employ threat displays that vary from the comic, in the case of the shingleback sticking out its bright blue tongue, to the dramatic, with the exploding neck ruff of the frilled lizard. There is also the bizarre display of *Moloch horridus*, the thorny devil, who simply hides his head between his armoured forelimbs and hopes his harlequin-patterned spiked body will scare off intruders.

Most of these lizards belong to the group known as agamids, which are known in Australia as dragons because of their spiny bodies and aggressive threat displays. They range in size from a few centimetres to about one metre and they continually move from shade to sun to shade to keep their cold-blooded bodies at a tolerable temperature. Their diet consists mainly of insects, which provide them with food and moisture. A thorny devil might polish off over a thousand ants in a day to ensure it gets an adequate intake of both. At least one minute nocturnal gecko is capable of living a year on the fats and moisture stored in its body.

The wastes of the Simpson Desert attract the black-faced wood swallow and lively zebra finch. The insect-eating wood swallow gets its nutrition and moisture from its prey and can survive for long periods without water, but the little seed-eating zebra finch needs water at least once a day. However, the finches can do without surface water for long periods when there is dew about. They are one of the few birds able to suck up the dew instead of scooping and swallowing with their heads held high.

The desert fringes provide a barren but secure haven for increasing herds of feral camels, donkeys and brumbies. They have shown skilled adaptation to the rigours of desert life and their wide-ranging habits have enabled them to find reliable water sources over large areas. The camels in particular have thrived since they first fled to the desert in the 1800s, and nowadays are in great demand by Arab sheikhs who find them superb steeds for camel races. Isolated in the Australian interior, they have avoided a plague of diseases that have ravaged the inbred camel herds of the Middle East in recent decades.

Curious feral camels (ABOVE) lope along, secure in their adopted environment, while an old male (BELOW RIGHT) surveys the world during a rest in the soft sand. Nearby a relaxed but wary dingo (BELOW LEFT) has a scratch.
OPPOSITE: Donkeys (BELOW) are another species that have successfully adapted to the desert. A languid desert goanna (LEFT) revels in the rare opportunity of having an afternoon soak in a rain puddle.

A white-plumed honeyeater (ABOVE) peers out between the spiky, moisture-retaining leaves of a
desert tree, while a well-armoured antlion larva (BELOW) emerges from the soil.
OPPOSITE: (Top to bottom.) When the rains come, the diminutive zebra finch (LEFT) immediately
begins his courtship dance. A shy topknot pigeon (RIGHT) prepares for flight. A dtella gecko (LEFT)
and a desert skink (RIGHT) settle down on tree branches. A lean and tough bulldog ant (LEFT) leaves
its underground nest to seek prey, while nearby, an equally tough black beetle (RIGHT) scavenges.

The arid interior's waterholes are also shared by dingoes and in times of plenty after rare rains, they venture into the heart of the Simpson Desert to hunt, finally returning to their old haunts in a wasted body frame when the pickings are exhausted. Bold and curious creatures, they show no fear of man in this wild environment.

The eastern dunes are spaced by long, skinny ephemeral lakes with poetic, tongue-twisting Aboriginal names – Mirranpongapongunna, Peera Peera Poolanna, Peedeemoondinna, Pompapillina, Pirriepatchillie... The names celebrate events in a long history: life-saving rains, good hunting, the desert blooming, and death and disaster. They also celebrate the unique call of the desert.

Its spell has lured explorers since the early 1820s. Their names live on in a proud roll call – Charles Sturt, Edward Eyre, Ludwig Leichhardt, John McDouall Stuart, Robert O'Hara Burke, William Wills and Ernest Giles. Some achieved their goals, others succumbed, but all acknowledged the powerful lure of the so-called dead heart of Australia. It has been summed up succinctly by writer Ray Ericksen, who retraced the tracks of Giles, the most energetic explorer of the Simpson Desert. He explained the call in these words: 'I missed the hard forms, the contorted branches and the peeling bark; and I missed the colour, the great range of muddy greens, the fiery reds and the laughing yellows. I missed the unpredictability of it all: the smell of eucalyptus oil drawn off in vapour by the sun, the scent of boronia heavy in the evening air. I missed the sounds: twigs and leaves swish-rattling along the ground before the blast of a hot northerly. I wanted a hard light again and a genuinely blue sky.'

Seashell relics (BELOW) *of an ancient sea pebble many pans.*
OPPOSITE: *A white-trunked gum* (TOP LEFT) *shows the scars of desert life, while the rains came too late for this tree* (TOP RIGHT). *The dry river beds host a healthy variety of plant life through wet and dry* (BELOW).

We are camped at Oodnadatta, named after the spiky yellow flower of the mulga, when we hear that the rains have come to the desert for the first time in nine years. We fly east to a long-abandoned airstrip near Poolowanna Lake on the western edge of the ephemeral lakes. Poolowanna is dry, but the soil is studded with dusty craters from raindrops that have already sunk into the thirsty soil. We set out for the nearest dune. Nothing seems changed, everything looks as desolate as before. A cool wind that smells of rain is whipping the sand off the red dune underfoot.

Shadows lengthen. We rest in the shade of a dune. At our feet we suddenly notice little shadows everywhere in the baked soil. Closer investigation shows thousands of one-centimetre high shoots, perhaps five or six species. On hands and knees we search and find seeds and shoots right across the valley. They range in size from the equivalent of alfalfa seeds to the round ball-shape of mung beans. The few seeds that failed to germinate are speckled and hard-cased. They taste bitter.

Elated, we finally set off for Birdsville to the north-east. Soon we are in a flooded landscape. Every depression glitters with water. The nearer we get to Birdsville, the deeper the lakes appear. The crests of the sand dunes are red-frothed in a sea of water. Barely 75 mm of rain had transformed the parched soil. From past recorded observations of this rare phenomenon, the flooding of the Simpson Desert, we know that in immediate response the black-faced wood swallow will be courting, that the zebra finch will be busy seeking nesting material, and that the galahs, red-tailed black cockatoos and green budgerigars will be leaving their regular waterholes to take part in the feast that will soon green the desert. In a day, a host of insects will hatch to add to the banquet for the birds and lizards.

The arid-zone flora have also adapted to withstand the heat and cold, harsh desert winds and long droughts, and to take advantage of the wet. Their seeds lie buried and dormant for years. They are coated with growth inhibitors – water-soluble chemicals – that prevent the seeds from reacting to minor showers. Only when they are dissolved by enough rain to sustain them through their life cycle do they germinate. Then the response is immediate and the fast-growing annuals complete their life cycle in just a few weeks.

A Darling lily (ABOVE LEFT) sparkles with new growth and a young brolga (ABOVE RIGHT) practises a dance step in antici-pation of the breeding season. Red gibber stones (BELOW) contrast with the fresh green growth brought on by the unseasonal rains.
OPPOSITE: *When the rains come, it does not take long for green life to surge into being. Top, left to right – a single drop of rain was probably enough to sprout this hardy plant; a cats head glows and a seed's new roots dig in. Bottom, left to right – a copper burr, a hardy perennial and a saltbush forge through the cracked clay of the dried out pans.*

The last light of the day highlights the crest of a red sand dune (ABOVE). The paler dunes to the south (RIGHT) meander across the flat landscape.

Even the fruits of the desert are attractive. The sandhill spider flower's fruit (ABOVE) contrasts with the fruit of the annual saltbush (ABOVE LEFT). The potato family produces a rich variety of flowers like the desert nightshade (BELOW). A black ant investigates a bluebush pea flower (BELOW LEFT).

OPPOSITE: *The Lechenaultia divaricata (ABOVE) and silvertail (BELOW) are typical desert flowers.*

A few weeks later. The red dunes are isolated by a sea of colour. Huge swathes of purple parakeelya carpet the sand-plains and the sand-dune flanks. Fluffy mulla mulla with white, pink and yellow flowers crowd together on their bushy clumps. Blue pincushion blooms and yellow-and-white poached-egg daisies keep the spinifex company. Pink and golden everlastings crowd the edges of claypans, while the beautiful Sturt's desert rose blooms on stony slopes to the north-east.

Noisy grey and pink galahs, swarms of chirpy budgerigars and the first of the resourceful grey teal gather near the bigger lakes. There is a noisy, babbling chorus of greed as they celebrate the virility of the desert in bloom.

It is, however, a temporary Eden. The harsh and unrelenting sun sucks the moisture out of the riverbeds and pans, and all too soon the desert reclaims its own. Within a month or two most of the pans are dried out, the rivers having exhausted themselves within days and only the larger lakes retain moisture for a season or two. Few will have any moisture left by the time of the next big wet.

Still, each morning the signatures of nature will be there on the dunes ... the busy tracks of the creatures of the night who thrive in the living desert.

A white ibis (LEFT) takes to the air, while an Australian bustard (BELOW) keeps a weather eye on proceedings.
OPPOSITE: *The elegance of flight is aptly demonstrated by two long-necked herons (TOP), a noisy flock of galahs (CENTRE) and a floppy-winged whistling kite (BOTTOM).*

With long pink legs strung out behind them, a flock of pied stilts (ABOVE) heads for a nearby waterhole. Screeching little corellas (BELOW) take flight. The end of the day (RIGHT) is treated to a waning moon.

Waves of forested hills flow to infinity from the flanks of Mount Bogong in north-eastern Victoria.
FOLLOWING PAGES: *Hell's Gap sinks away from a scarred hillside on the northern approach to Mount Bogong.*

THE LOST WILDERNESS
Bogong National Park, Victoria

A thick smoke haze, the legacy of Victoria's worst bushfires this century, shades the sinking late afternoon sun, while the impact of intensive logging gives the hilltops in the middle distance a scraggly appearance. A wary kangaroo (RIGHT) surveys its diminishing domain.
FOLLOWING PAGES: *Verdant growth in one of Victoria's few remaining rainforests.*

LOCATION: *North-west Victoria.*
SIZE: *144 000 hectares.*
GEOLOGY: *Varied origins with highly eroded tablelands the main feature.*
CLIMATE: *Rainfall of up to 2400 mm in Mount Bogong area with large rain-shadow areas to the east.*
FLORA: *Characteristically delicate alpine herbs and shrubs.*
FAUNA: *Home of the rare mountain pigmy possum, 146 bird species and numbers of introduced species including Sambar deer, brumbies, foxes and hares.*

BOGONG NATIONAL PARK, VICTORIA

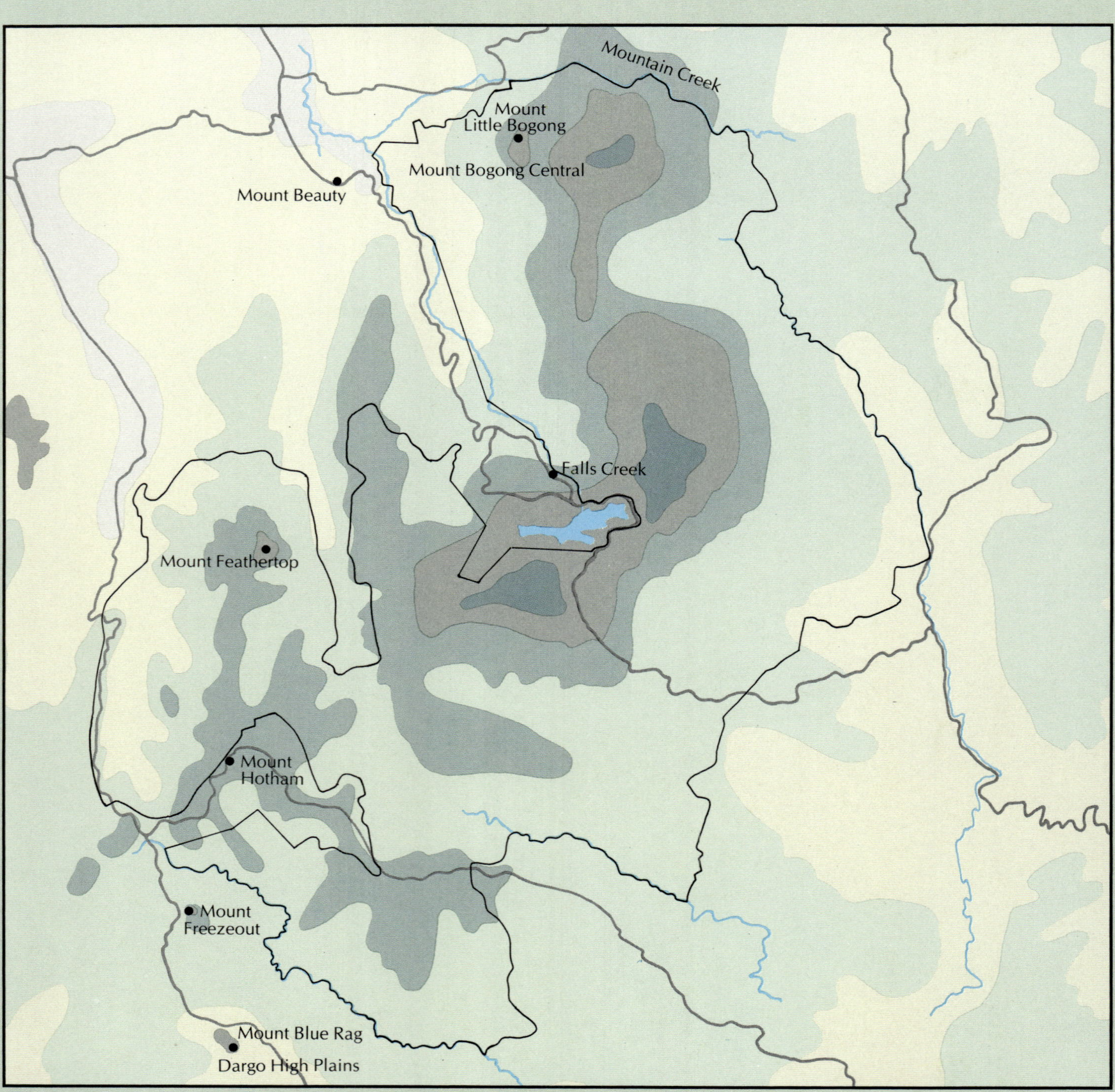

Mountain Creek

Mount
Little Bogong

Mount Bogong Central

Mount Beauty

Falls Creek

Mount Feathertop

Mount
Hotham

Mount
Freezeout

Mount Blue Rag

Dargo High Plains

THE COOLER SOUTH-EASTERLY wind that had finally brought relief from the dust storms and fire-fuelled clouds of smoke died down in the black hours before dawn. The haze that had blocked the stars for many nights was finally gone; a new moon crept up over the ridge opposite camp. Suddenly, the bay of a half-breed dingo rose from the valley. It was a cry of relief, a lifting of spirit... it was also confused and distorted. It was the cry of a wild creature trapped in an alien body and at that instant, its mournful cry seemed to symbolise the wilderness dilemma in Australia's most densely populated state.

Victoria, sadly, has only a few remaining pockets of wilderness. Even the vast State Forests that carpet the mountainous north-eastern half have been tamed. Below that canopy, a benign scenario is played. Fat Hereford cattle graze the undergrowth, methodically mowing each season's new growth. Like biological lawn-movers they nip shrubs and trees in the bud and eliminate everything except resilient grasses and persistent weeds. These hordes of cattle, like fire, serve as a force in altering the landscape. Their steady munching has virtually eliminated any juvenile tree growth and what remains is a generation of mature trees approaching senility.

In tandem with the grazing goes the felling of the forest giants leaving bare patches of giant skittles – discarded logs felled to ease access to the choice lumber.

All this activity has caused the widespread introduction of exotic weeds, particularly the ubiquitous blackberry. This notorious invader favours any bare patch of soil and the exposed banks of streams and rivers. It soon creates an impassable barrier and establishes a favourable habitat for feral cats, foxes and rabbits. The succulent berries entice the diminishing bird life within easy range of fox and cat.

Our original destination was the rugged Mount Cobberas region of north-east Victoria. A two-day search failed to reveal anything that even remotely resembled wilderness. The whole area had succumbed to grazing, logging, weeds and erosion – a 'wilderness' only, perhaps, in the bleak and barren sense portrayed in the *Bible*.

Fire and logging, the major threats to the world's shrinking forests, have decimated vast areas of Victoria. A logged slope, also victim of fire, provides a stark foreground to the original forest seen behind.

We continued our search. At Omeo we spoke to the local agent in the Lands Department office. He recommended that we go to an area north of Dargo. 'It's full of useless country,' he said. His sky-blue eyes, used to scanning distant horizons for smoke and noxious weeds, were serious. He was right. It was useless country – a repeat of the Mount Cobberas experience.

The search continued, the mountains of the Victorian Alps luring us north. They looked promising as we tackled the winding road from Mount Hotham via Mount Beauty and on to Falls Creek – the core of Bogong National Park. Nightfall saw us camped next to an aqueduct feeding the Pretty Plains Dam, part of the hydro-electricity scheme that has mutilated the alpine plateau with power lines, dams, roads, pipelines, concrete blockhouses and signs. Signs that read 'No camping', 'No Trespassing', 'No driving on verge' were originally planted to protect the delicate alpine plant life that is now being trodden underfoot by sleek cattle. Sadly, the power system need not have been built as the much more ambitious Snowy Mountains Scheme soon followed and its output was more than adequate for the eastern seaboard's demands for power.

Exploring the area, we found little unaltered landscape. The rolling alpine meadows were freckled with the brown-and-white Hereford cattle. Ski-pole lines marched from horizon to horizon; fire trails and the main highway between Omeo and Mount Beauty meandered along contour lines and ridges. Sunset saw the sad spectacle of introduced trout rising to feed on flies in the narrow, truncated and slimy aqueduct. We broke camp at first light.

Later a local newspaper informed us that there are 20 000 head of cattle worth $3 500 000 grazing the high plains. The mountain men are due soon for the annual muster and apparently are not happy with threatened reductions in grazing leases. Government policy is to reduce the available leases by 10 per cent initially, and gradually to withdraw all grazing leases. It is an emotive issue. A way of life will be destroyed, traditions eradicated, folk heroes (like The Man from Snowy River) will retreat further into history and 20 000 head of cattle will have to chew the cud elsewhere. In a drought year, it is not easy to argue about the long-term regeneration plans for the area.

An axed tree trunk (LEFT) *inscribed with deep foot-wedge holes stands like a headstone. It is framed by the ubiquitous blackberry and an introduced thistle holds court in the foreground.*
A skew-horned Hereford bull (BELOW) *chews the cud oblivious to the damage it inflicts on its surroundings. A fallen forest giant* (RIGHT), *a victim of sheer old age, lies gracefully on a bed of leaves.*

Narrow contour lines on a large-scale map of an area to the south lure us with the promise of steep gorges. The four-wheel drive track takes us past the evocatively named Mount Freezeout – a basalt pimple in reality. We head on down Blue Rag Track and along Basalt Ridge. Our spirits rise. The area has barely been logged and the various stages in growth of the trees indicate grazing has been minimal. The rapid taper of the ridges has probably discouraged logging and grazing. We are eager to establish a base camp and record our impression of a wilderness oasis, to show and describe the peace and harmony found in the areas we have visited in our search for Australia's wilderness.

Basalt Knob looms up and we continue along an undulating ridge. The map shows contour lines nearly joining on the far side. We slow down. A sudden gasp. The gully ahead looks like a trendy outdoor restaurant with tables made of log stumps. That is all that remains of the forest behemoths that favoured the moistness of the gutted valley. The demolition continues further down the slope to reveal the mess left from logging. Bashed and battered reject logs lie in chaos, their sawn-through trunks mirrored, ring for ring, on the stumps. It is an open book of the area's history. The rings tell of drought and fire and years of plenty by their density, colour and thickness.

The silence in the clearing is eerie. No birds call. No insects buzz. The bleak and terrible landscape contrasts with the green and verdant scene offered by a far ridge. The enormous wastage – abandoned logs, the shattered stems of juvenile growth flattened by the falling trees, discarded loppings – ends as suddenly as it started. The track we follow has become a rutted highway for the forestry trucks.

The desolation brings to mind Aldo Leopold's observation in *A Sand Country Almanac*. He wrote: 'Every woodland, in addition to yielding lumber, fuel and posts, should provide its owner a liberal education. This crop of wisdom never fails, but it is not always harvested'.

The bold Victorian Alps landscape has been brutally scarred by human intrusion. The winding road leading to Mount Hotham (BELOW) has left eroded gullies and landslides in its wake. Other blatant intrusions are the omnipresent high tension power lines (RIGHT) that jackboot their way across the countryside, clearing all from their path.

The journey has barely begun it seems. We contact the Australian Conservation Foundation, the umbrella organisation for Australia's conservation and wilderness societies. 'Victoria is a bit difficult,' explained a senior spokesman. 'Even the deserts to the west are fairly tame.' He then explained that the areas we had already visited were considered viable by Victorian standards. 'You see, we have to be believers here in regenerative wilderness, and the areas you've seen qualify.' Further discussion followed which suggested we visit some isolated valleys north and east of Mount Bogong. 'They're only token wildernesses, but we consider them jewels. Look after them.'

We found the treasured areas – a series of remnant rainforest valleys protected from human intrusion by their sheer ruggedness – but only after wending our way through a minefield of weeds (including tobacco, blackberries and thistles), rusting car hulks, abandoned smallholdings and roads which gradually deteriorated into twin ruts.

Rural sentiment is basic and straightforward as these bumper stickers testify. But desolate despair is all that surfaces from drought-stricken scenes (BELOW) where flocks of sheep have shaved every living blade of grass. Logs left to rot after logging (RIGHT) are testimony to the waste the nature of the industry encourages.

Lush temperate rainforests survive in a few remote valleys in northern Victoria (LEFT) and provide refuge for a host of resplendent insects. Here moths, beetles, grasshoppers and worms live in delicate balance.

A yellow-tailed black cockatoo (ABOVE) *feeds in the forest canopy. Long shards of candlebark hang from the smooth trunks of tall alpine ash* (RIGHT) *before falling and adding their bulk to the layer of mulch that carpets the forest floor.*

A high-pitched 'coo' reveals the presence of a stately wonga pigeon on a eucalypt branch. Typically it squats motionless, facing away, with its head turned towards the intruder. Its presence is a sure sign of a peaceful environment as the wonga rarely tolerates human invasion of its territory.

Giant alpine ash soar skywards. Their smooth trunks, as straight and tapered as billiard cues, contrast cleanly with the luxuriant undergrowth of tree ferns, shrubs and juvenile trees.

The valley is an insect haven. Their presence offers a muted chorus of buzzes, whirrs, drones, chirps – an aural offering matched only by the counterpoint song of the birds. The forest floor, a rich layer of detritus, harbours a living tapestry of termites, beetles, ants, worms, grubs and moths. A single overturned log reveals a thriving wilderness in miniature. It is populated by brown and silver centipedes, translucent worms, sluggish spiders, armoured beetles and palettes of pastel fungi all co-operating in a symbiotic relationship to return a fallen tree to the soil – as soil. There is nothing wasted, nothing lost in the orchestrated cycle of ecological balance. It is stable, secure, complete.

Shards of candlebark peeling off the alpine ash add to the forest chorus as they fall to earth. The fertile mulch is boosted by the noisy activities of yellow-tailed black cockatoos ripping bark off acacias and eucalypts in search of 'white' grubs. Their feasting is a squawky, crabby and petulant affair, a surprising contrast to their relaxed and buoyant flight, an aerial ballet of effortless and graceful movements with broad wings spread and long tail fanned out.

An immature crimson rosella (ABOVE) cocks its head for a quizzical stare. Curious eastern yellow robins (RIGHT) also keep a close watch on forest proceedings, which include hosts of yellow butterflies feeding on nectar (BELOW).

One cool morning, an aural blast shatters the forest calm. The gamut of whistles, chirps, song and squawks being broadcast from nearby undergrowth sounds like the feathered denizens of the valley had gathered for an eisteddfod. However, closer inspection reveals the passionate bard of the bush, the Superb lyrebird, in full cry to charm a shy and reticent female of the species.

His mimicry, combined with a dramatic and complex dance, finally settles down to a display of his elegant white lyre-shaped tail and a persistent 'cric-cric-cric' call which culminates in mating.

White-eared honeyeaters boldly forage nearby. Their mellow call 'cherryweet, cherryweet' persists. It contrasts sharply with the whipcrack call of the eastern whipbirds bouncing briskly through the undergrowth. The gregarious and curious silvereye myopically investigates camp activities and hustles for scraps. Another regular visitor is a rufous fantail, which looks like a darting flame in the forest shadows. It playfully performs free-falling acrobatics not unlike an unbalanced shuttlecock as it hunts its elusive insect prey.

Two fallen forest giants form a 'V' that entice us down to the creek that courses through the valley. We use one mossy log as a bridge across the dense undergrowth. The 1.5 metre thick trunk is soft underfoot, a sponge created by an army of insects devoted to the decades-long task of chewing it up. The creek gurgles cheerfully below, its low summer level refreshed during the night by a long rain shower. It is flanked by a thick forest of tree ferns, friendly shelters for the many insect-eating passeriforms thriving there. Their shady canopies also prevent the growth of light-loving weeds.

Higher up the valley the tree ferns gradually give way to the hard-leafed sclerophyll forest dominated by eucalypts bearded with moss. The tree-line is populated with gnarled snow gums with knuckles creased from battling the elements in their exposed eyries.

Undisturbed rainforest (BELOW AND RIGHT) *provide a tranquil setting for nature's chain of life. Candlebark, dying fronds of tree fern, fallen leaves and logs are constantly recycled by termites, worms and beetles.*

The bright green hues of rainforest plants and ferns (ABOVE AND BELOW) are offset by the paler shades of grey and green of the alpine ridges above the snowline (RIGHT).

The creased knuckle of a snow gum (LEFT) reflects the tough terrain that it thrives in. On exposed slopes,
the snow gums bend with the prevailing wind (ABOVE) while in sheltered gullies (BELOW) they strive upwards.

Most of the flowers found in the alpine regions are tiny, but tough, as typified by the bearded heath carpet (ABOVE). A pale mint bush (TOP LEFT) has its colours reflected in the fruit of a lillypilly tree (TOP RIGHT). Members of the Olearia family (LEFT AND BELOW) can be found in many valleys, while the flower of this herbaceous shrub (Arthropodium species) (RIGHT) is fairly rare.

Australia's evergreen forests had their historical counterparts in the Mediterranean lands. There, when fire and man had cleared the vast tracts, goats went to work and have remained a dominant factor. The devastation set in early and was observed at the height of Athenian glory by philosopher Plato, who noted: 'What now remains compared with what then existed is like the skeleton of a sick man, all the fat and soft earth having been wasted away, and only the bare framework of the land being left'.

The valley we found was a rare survivor in a tamed environment. Plato's observations would be as valid today over most of Australia. It is time to heed the reminder given by naturalist Marston Bates: 'Man's destiny is tied to nature's destiny and the arrogance of the engineering mind does not change this. Man may be a very peculiar animal, but he is still part of the system of nature'.

The system of nature abounds with examples of symbiotic relationships, mutualism, balanced biospheres and environmental harmony. It can be as simple as the world of lichens. Each lichen 'plant' is made up of a fungus and an alga growing in a close relationship, and neither partner in the union is found alone in nature. The fungus provides the 'skeleton' in which the single-celled algae grow and absorb water and salts from the environment. The algae, through photosynthesis, build up organic food for both partners.

Lichens are often found on exposed rock outcrops and the north-east slope of Mount Bogong is no exception. Here a knife-edge of granite called Hell's Gap welds the mountain to a long, undulating ridge that spreads manicured fingers of logged forest scarps into the foothills. Mount Bogong itself is no longer grazed and Hell's Gap provides a natural barrier against further intrusions.

The slopes of Mount Bogong provide refuge for clumps of snow daisies (BELOW). Many other blooms (OPPOSITE) are found nearby. From top, left to right, they include verbascum (an introduced weed), a striped gentian, a spiky acaena, yellow paper daisies, a vigorous old man's beard (clematis) and clusters of minute alpine daisies ready to open with the sun's first rays.

Hell's Gap is our route to the rounded crest of Mount Bogong. Plump clouds squat low in the valleys. A light wind rising from the broad valleys to the north-west keeps a family of crested hawks aloft along a ridge below us. They peel off at irregular intervals to plummet earthwards, talons outstretched. One catches a small animal and devours it on the wing.

The serious business of hunting is often interrupted by a solo hawk soaring several hundred metres high and then tumbling in an apparently unco-ordinated freefall. Ornithologists may have theories for this behaviour, but to us, the tumbling falls are play, a celebration of another day in the wild.

Dawn sees a mist surround camp. It revives memories of another morning on another visit to these mountains...a winter morning. An overnight fall of snow had powdered the Bogong High Plains. The cosmetic covering hid the cross-country ski tracks of the day before and aqueducts, roads, pipelines, cattle trails and dung had disappeared under 30 centimetres of fresh snow.

Across the Big River valley Mount Bogong lay hidden under cloud, a white skirt fringing her flanks. To the south the blanket of white continued and for an hour or so the original outline of Victorian wilderness lived again.

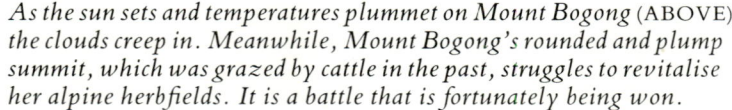

As the sun sets and temperatures plummet on Mount Bogong (ABOVE)
*the clouds creep in. Meanwhile, Mount Bogong's rounded and plump
summit, which was grazed by cattle in the past, struggles to revitalise
her alpine herbfields. It is a battle that is fortunately being won.*

A rocky sentinel stands guard near the summit of the Ironbound Range, while the Louisa River snakes across the valley below.
FOLLOWING PAGES: *The bleak beaches fringing Wilson Bight shelter beneath the weather-worn slopes of the distant Amy Range.*

THE LAST WILDERNESS
Southwest National Park, Tasmania

Heavy seas break long before reaching the rocky shores of Wilson Bight.

LOCATION: *South-west Tasmania.*
SIZE: *Over 1 000 000 hectares.*
GEOLOGY: *Mainly pre-Cambrian sedimentary rock up to 1000 million years old.*
CLIMATE: *Rainfall of over 3000 mm influenced by predominantly westerly trade winds.*
FLORA: *Originally all rainforest, but bush fires have allowed eucalypts to take over most areas with exception of wetlands.*
FAUNA: *Extremely diverse and ancient species of invertebrates, mammals, amphibia and crustaceae. Birdlife plentiful with many migrant seabirds.*

SOUTHWEST NATIONAL PARK, TASMANIA

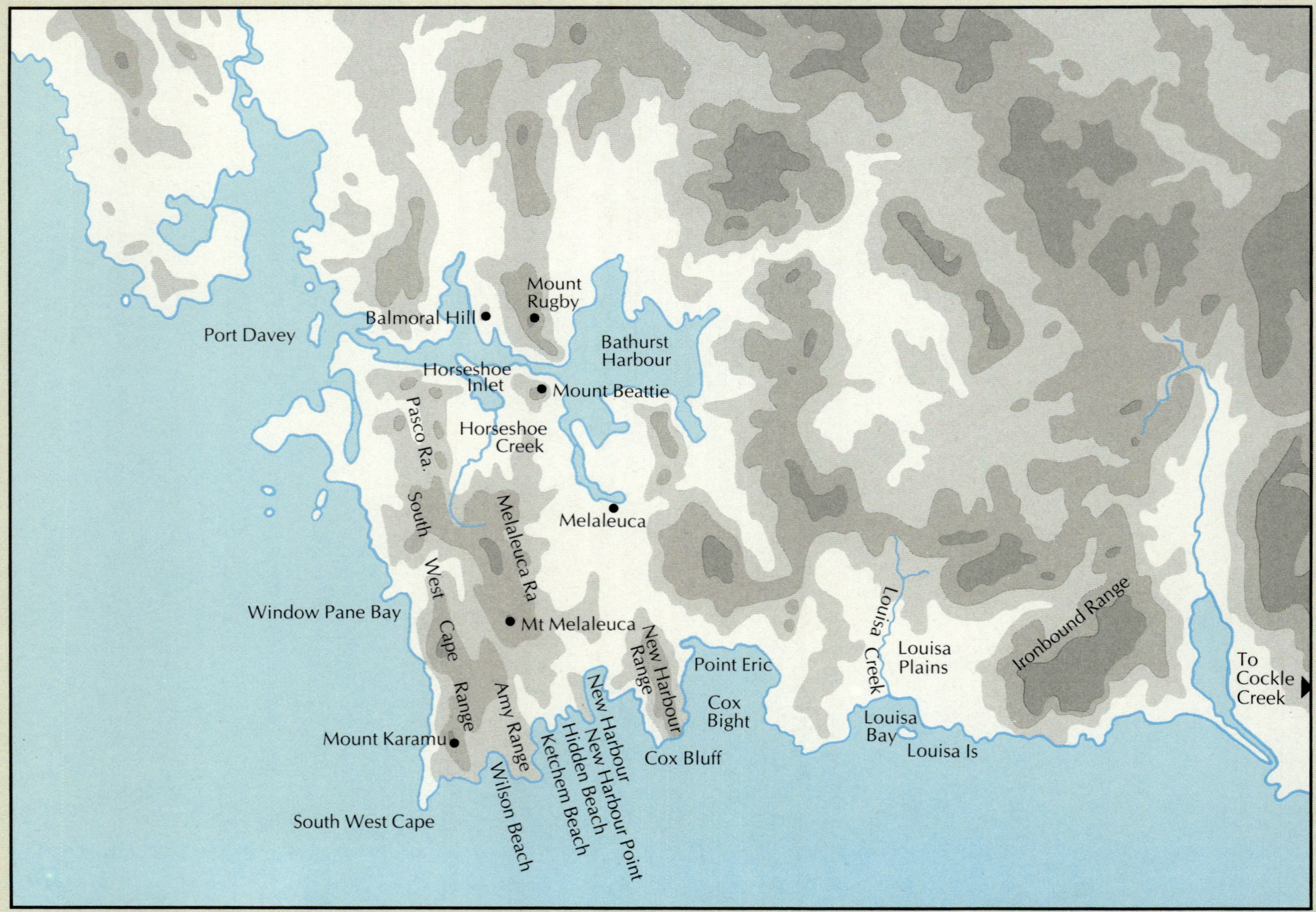

Port Davey

Balmoral Hill •

Mount Rugby •

Bathurst Harbour

Horseshoe Inlet

• Mount Beattie

Horseshoe Creek

Pasco Ra.

South West Cape Range

Melaleuca Ra.

Melaleuca •

• Mt Melaleuca

Window Pane Bay

New Harbour Range

Point Eric

Cox Bight

Louisa Creek

Louisa Plains

Ironbound Range

To Cockle Creek

New Harbour Point
New Harbour
Hidden Beach
Ketchem Beach

Cox Bluff

Louisa Bay

Louisa Is

Mount Karamu •

Amy Range

Wilson Beach

South West Cape

THE WHITE QUARTZ flank of New Harbour Point juts through the wild breakers stretching a kilometre out to sea. A setting sun gilds its reflective canvas, throwing it into cheerful contrast with the grey hulk of De Witt Island and the bleak swells queuing up for their death dance on Hidden Beach. For nine days the sun has hibernated deep in cloud cover and now it sweeps across the storm-washed beach and out to sea over the frothing wave-crests which have welded together in a turmoil of foam. On the beach the noise is numbing, not unlike the roar of a steam locomotive at full bore – going nowhere.

The wild surf has been battering the southern shores of Tasmania for weeks following a rapid sequence of gales in the bleak southern oceans. It is unrelenting in its savage pounding, its constant sacrifice of fresh waves. The rugged cliffs and fingers of land that face this rampant attack have seen it all before and show their age. They are all battle scarred: cliff bases are a confusion of dolerite and sandstone; the rocky headlands that bear the brunt of the attack are scoured down to base rock, but they have done their duty and protected the series of scalloped bays and inlets that rim the coast.

The creek that twists across Hidden Beach is swollen by the heavy rain which has been falling intermittently for a week and it is the colour of strong billy tea, a characteristic hue acquired during its downhill rush through the inland peat swamps and buttongrass plains. It mixes with the surf and creates a giant chocolate milkshake. An enormous sand blow dominates the curve of the beach and its leeside is a thick forest, a dark green mass broken only by the weathered skeletal remains of fire-scalded forest giants poking through the canopy.

A cold southerly wind comes in from Antarctic waters. In a forest clearing our driftwood camp fire blazes. The birds of the forest are in full chorus as they prepare to settle for the night. There is calm purpose in their calls – no sense of urgency or threat, just gentle probing chirps and soft affirmations. Just the end of another day in the last Australian wilderness...

Wave after wave of rhythmic swells file into Surprise Bay, while cloud covers the ominous folds of the Ironbound Range which in turn hides the craggy outcrop of New Harbour Point to the west.

Photography by Allan Moult

A sooty oystercatcher (ABOVE) fossicks along the turbulent shoreline of Wilson Beach during a brief sunny interlude. For days in a row, heavy mists settle on the beaches of Wilson Bight (BELOW AND RIGHT). They come without warning at any time of the year.

Peat-soaked creeks (LEFT) *take on a brown tinge in their journey to the sea* (ABOVE) *where they stain the white sands with their distinctive hue.*

Morning brings a clear sky and a warmer wind from the west. The stream is gurgling with diminishing urgency, the surf's roar has softened into a steady bass rhythm and there is staccato ad-libbing by the birds. Sphagnum moss is a soft springy carpet underfoot. Laurel leaves provide an oily catalyst for our fire. Their lusty blaze soon ignites the myrtle twigs. Dry sassafras and melaleuca leaves add their special aroma as the salt-dusted driftwood blazes. Just another morning…

The Southwest National Park is a jig-saw bordered area of over one million hectares enclosing the jewel that is, perhaps, the last Australian wilderness. Its curly fringe is a compromise with mining and forestry interests, but the core is there and future expansion is possible.

The wildest sector, our target, is the south-west corner – a bleak, storm-lashed peninsula spiking the southern seas. Geologically it is ancient terrain, but in terms of human presence it is recent. While the Aboriginal inhabitants of Tasmania lived there for 20 000 years or more, they penetrated to the South-West perhaps less than 3000 years ago.

Sir Mark Oliphant, a former president of the Australian Conservation Foundation, puts their role in perspective: 'These decent, peaceful people lived in villages of huts, cremated their dead and fired the bush and forest to drive out animals for food, changing greatly the ecology of the region. Once plentiful and able to eke out a reasonable subsistence in that infertile region, they were wiped out by the white government, which herded them in settlements where they could become Christians and die of the white man's diseases.'

New Harbour Point (BELOW) *basks in the sun during a rare break in the weather. Spiky buttongrass plains* (ABOVE RIGHT) *mimic the skeletal remains of fire-ravaged trees jutting above the forest canopy. A tranquil setting* (BELOW RIGHT) *alongside the Hidden Bay lagoon.*

The Toogee Low tribe seasonally roamed the buttongrass plains that stretch from Cox Bight past Melaleuca to Bathurst Harbour, an inner part of Port Davey, which enabled them to avoid the rigours of the exposed south-western corner of Tasmania.

The first white man to follow this path was George Augustus Robinson, in 1830. His journals record the myth of one of the Toogee devils, an evil spirit called Wraggeowrapper who was, they said 'like a black man only very big and ugly... he travels like the wind, he comes and watches the natives all night and before daylight comes he goes away like swift wind'. He too was watching us on our first night...

We fly from Hobart on a rainy summer afternoon, our destination Cox Bight, for a food drop, and then on to Melaleuca, five minutes flying to the north. Within minutes of take-off the weather closes in and the pilot is forced to head through a gully below cloud level to the coast and then hug the beaches at low level. Occasional breaks in the storm reveal the 1200 metre peak of Precipitous Bluff, which rises sheer less than two kilometres away from the coast. Further north other jagged peaks and mountain ranges pierce the cloud. The landing on Cox Bight is a skidding slide as we come in sharply angled to the wind. We secure our food drop in the fork of a tree and soon disembark at the bulldozed airstrip at Melaleuca.

Base for the night is the Charles King Memorial Hut, a personal tribute by his son, Denny, one of the legends of the South-West. For 50 years there has been a King at Melaleuca, and both father and son are renowned among bushwalkers throughout Australia for their hospitality and unstinted efforts to preserve the wilderness.

A cocky superb blue wren scurries round the hut snapping up food scraps. Overhead, nesting swallows zip in and out from their mud nest above the fireplace. Denny has cut a hole in the wall above the door to make their journeys easier. Sleep comes easily despite the wind whistling through the swallow's exit.

Hidden Bay (BELOW) *scallops the forest-fringed buttongrass plains that lead down to the coast from Melaleuca. The rocky foreshores of Hidden Bay* (RIGHT) *are typical of most of the beaches along this wild coastline.*

A freshly-hatched pied oystercatcher chick (ABOVE) lies immobile in its shallow nest hollow on a beach. The limp remains of a large jellyfish (BELOW), tossed ashore in a storm, decorate the high tidemark. The quartzite slopes of the Amy Range (RIGHT) dominate the eastern shores of Wilson Bight.

By noon we reach the vast buttongrass swamps, riddled with the burrow systems of the freshwater crayfish that buffer the southern shores of Horseshoe Inlet. We leave the Port Davey Track, which traces its muddy trail to the north where the cloud-capped view of Mount Rugby is framed by pointy Balmoral Hill and Mount Beattie. It is a bleak vista of diminishing greys. To the west, the Pasco Range, our gateway to the South West Cape, is also covered in cloud.

It rains all day. Our trail across the swamp is a hard muddy slog punctuated by booby traps of water-rat burrows and scrub bashing through thick tea-tree and melaleuca thickets lining Horseshoe Creek. Our boots gather a glutinous, primeval jelly coating not unlike transparent leeches. Our initial campsite was a high ledge, wet underfoot but seemingly protected by the bulk of a hill below the Pasco Range. The wind, however, proves fickle as it suddenly builds up and tears at the tent, lifting it high and threatening to sweep it up the valley and over the next range.

We flee to the comparative shelter of a patch of trees in a nearby hollow and pitch the tent on a bed of sphagnum moss. A fire is made on logs sunk into the morass. Everything is spongy underfoot. Stand still and you sink into the seep. There is constant groaning and creaking as the wind rubs trees together. Smooth channels of wear show it is an ongoing occurrence. We are under constant attack from leeches, but at least our hidden haven is safe from the brunt of westerlies. The wind is unrelenting. The spirit Wraggeowrapper is on watch. He stays all night and leaves before daylight. The wind does not.

Grass-covered layers of peat, striking rock formations and bristly grasses decorate the shores of the Hidden Bay lagoon (LEFT AND ABOVE) in a setting not unlike a tranquil Japanese garden. A wind-crippled leatherwood tree in full flower sits on one of the incongruous grass platforms that tide-mark the southern shores of Horseshoe Inlet (BELOW).

Strange, spongy fingers of lawn poke into Horseshoe Inlet like arthritic claws (ABOVE AND RIGHT). *Somehow the delicate blades of grass survive being dunked by the continual salt spray from the wind-tossed inlet.*

The Pasco Range is still under heavy cloud. Intermittently the sky clears and sun filters through the leafy canopy. It cheers us briefly until we note the speed of the clouds pacing overhead. We finally emerge from our steamy enclave and head west up a small hill to scout a route through the scrub. We are still uncertain about retreat. The wind decides. A sudden gust nearly fells us. (Later at Melaleuca we are informed that winds of 100 kilometres per hour have been recorded at nearby Bruny Island.) We survey the scrub below. Having battled through it the day before we are wary and elect to head for Horseshoe Inlet's forested foreshore.

We look for an easy way round its shores, perhaps wading its fringes. A long, wet slog follows across the swamp and we finally bash through 30 metres of tight scrub and reach a green lawn jutting out into the wind-frothed waters of the inlet. On both sides these strange spongy fingers of grass poke out like arthritic claws. Their base appears peat-like and the grass itself is as short as if had just been trimmed for Wimbledon.

The brackish, storm-tossed waters have helped keep the scrub in check, or at least thinned, for several metres inland and we hop along at full speed, only occasionally leaving our green stepping stones to bush-bash. A sharp bend and Horseshoe Creek appears – our nemesis from yesterday – its lee shore turning its peat-soaked, red-brown waters into a marble mirror, an impassable deep mirror. Beyond, a lone black swan takes flight after a long, ambling take-off into the face of the heavy wind. Wings beating hard it rises ponderously with its head low, but soon wearies and splashes down again.

Below the Pasco Range the wind has swivelled down in a confused melee. Yet higher, the clouds strut steadily on a fixed westerly course. When the sun does shine, the cloud shadows race across the plains and teasingly overtake us. There is no way we can keep pace, even if we run. We watch one large shadow speed its way across a plain of buttongrass. It reaches a tall hill and gropes up its fissured side even more swiftly, changing shape like a blob of jelly.

Ahead lies the swamp. We squelch across and half an hour later we emerge again on the South Coast Track. Three hours later, two bone-weary and tired trampers reach the hut at Melaleuca. In two days we have covered 36 kilometres in 17 hours, each of us with 32 kilogram packs!

Stringy seaweed patterned by the flow of the tides fringes Ketchem Bay (BELOW). New Harbour Point (ABOVE RIGHT) is framed by Ketchem Point and Ketchem Island. A large arch (BELOW RIGHT) dominates the western edge of the Amy Range as it settles into the Southern Ocean.

FOLLOWING PAGES: *Ketchem Beach and Ketchem Island dominate the foreground in this view to the east from the crest of the Amy Range.*

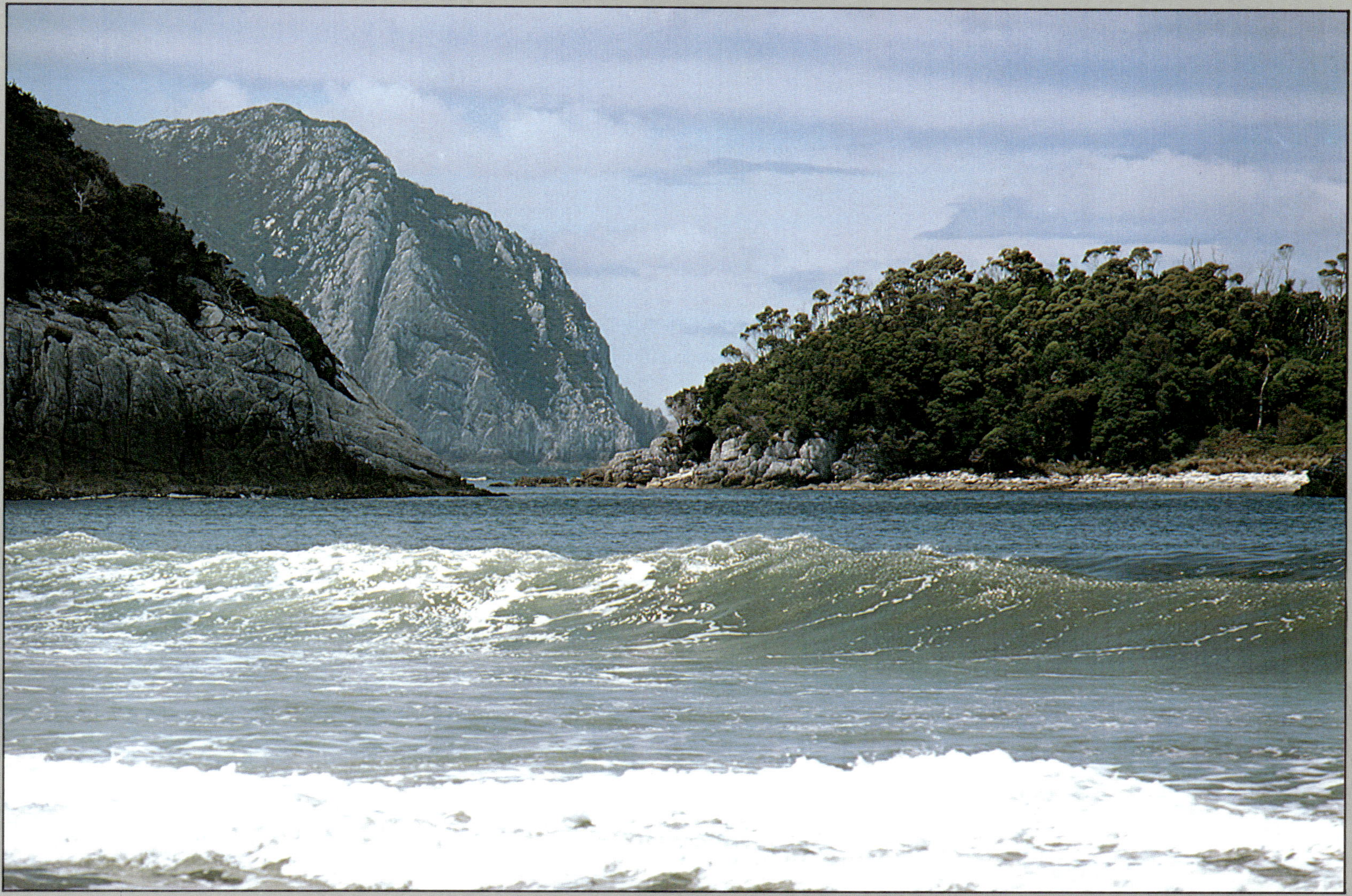

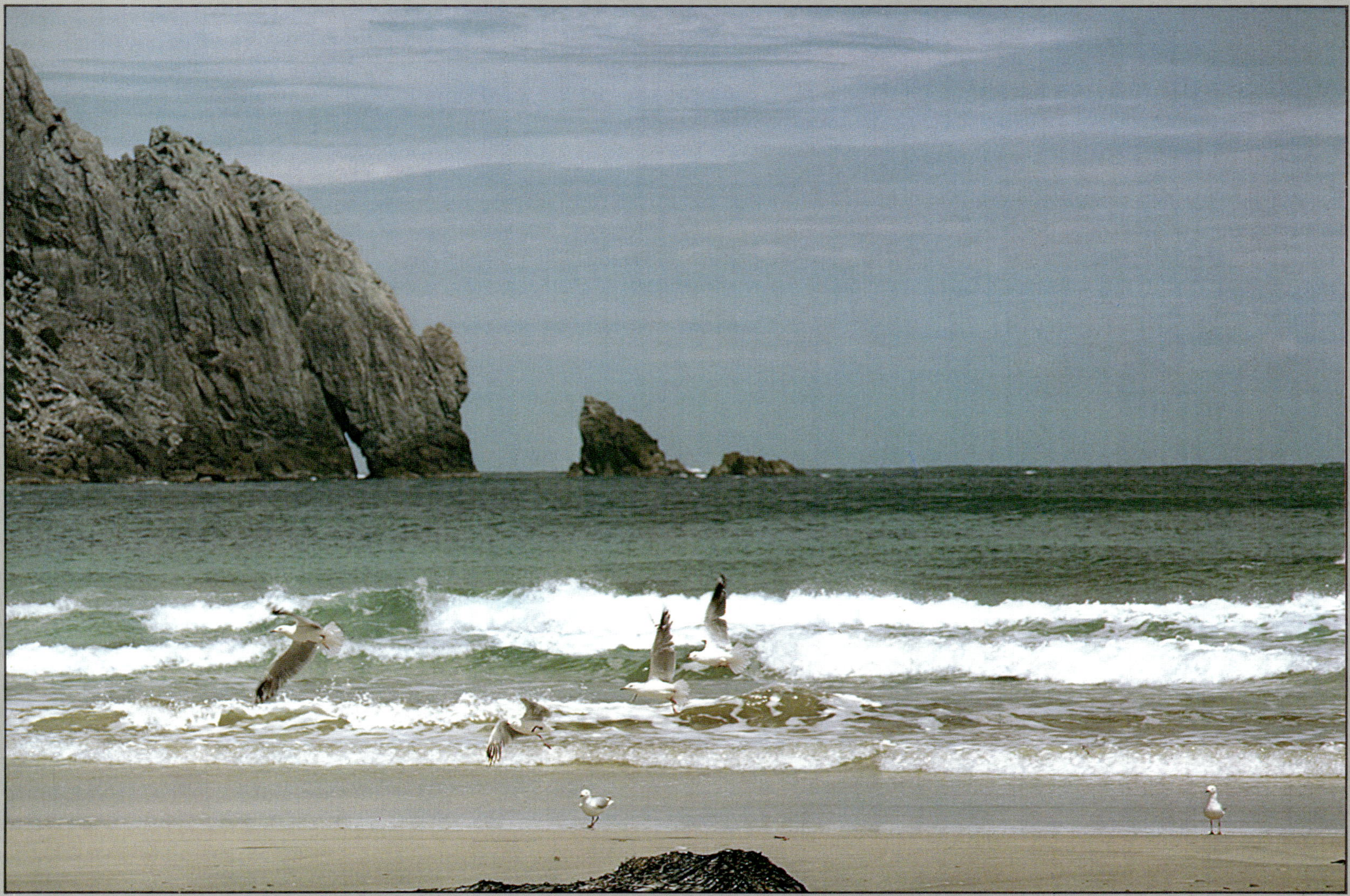

The journey continues south to New Harbour. It rains constantly. It is bleak and cold, not really spectacular, but full of a presence. New Harbour Point juts defiantly out to sea, a high wall protecting the anchorage from the devastation of the rugged westerlies. It sits serenely as storm clouds sweep overhead.

The beach is empty of footprints, the rain squalls have eroded all tracks. Out to sea, giant swells break in a froth of white on Outer Rocks. The Point is on duty. To the west is South West Cape. In between lies a series of parallel ranges and a string of beaches.

From New Harbour it is a stiff climb to the sentinel finger of New Harbour Point. This is where the trip begins in earnest. Elated we surge on. In this wild and wonderful coastal amphitheatre there is a magic in the air. The feeling persists. An awareness for the first time of being in wilderness – true wilderness – alone, but part of it, rooted in it.

Our campsite at Wilsons Beach is behind the sand dunes in a delightful clearing in rainforest. It has obviously been a base camp in the past for assaults on the South West Cape. Driftwood has been used for seating for a dozen souls round the fireplace. Firewood is neatly stacked for use. It appears the weather has kept more than one party locked in for long busy days.

To the west, Mount Karamu is hidden in cloud. It is our last beacon before the South West Cape. The next day is the same. The assault is aborted. Instead we explore the beach. The sun chooses to spotlight the sand through a gap in the heavy cloud cover. It is a long beach with sharp rocks standing like saw blades out to sea. Cracks and crevasses with hunks of rock and driftwood jammed fast four to five metres above high watermark bear testimony to violent storms.

Rainforest (BELOW) *dominates the coastal forests of south-west Tasmania. A low tide at Wilson Bight* (BELOW RIGHT) *leaves a sucker-footed kelp anchored high and dry* (ABOVE RIGHT).

The debris stranded on the beaches after one of the many violent storms that rage across the wild coastline of south-west
Tasmania is often a curious blend of the bizarre and the commonplace. It is also often a graveyard for exhausted birds (TOP LEFT),
jellyfish (BOTTOM RIGHT) and bluebottles (BOTTOM LEFT).
OPPOSITE: A crab's claw, a channel of water and a seaweed float nodule create temporary works of art on a sand canvas.

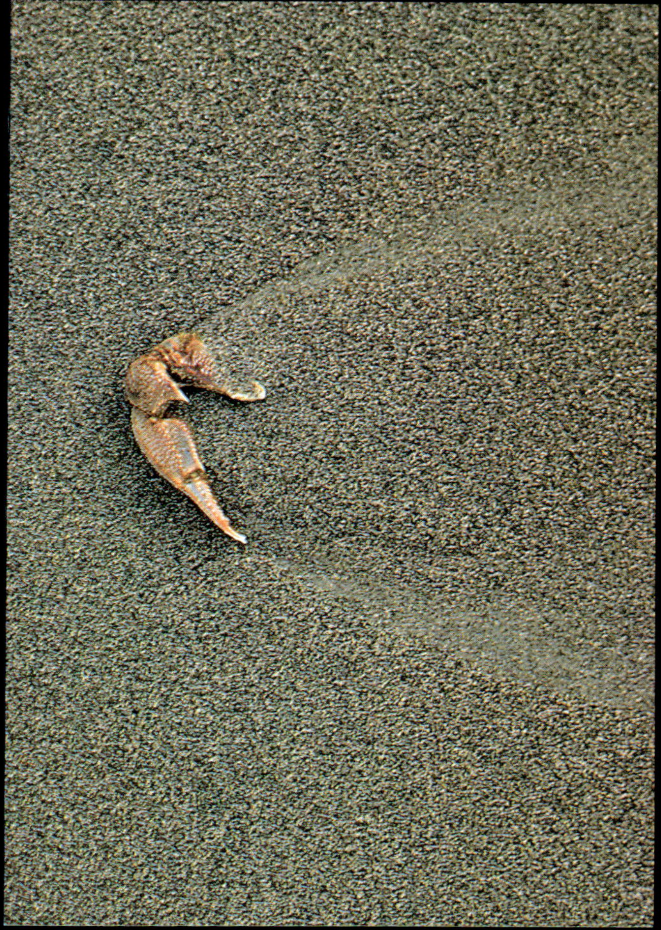
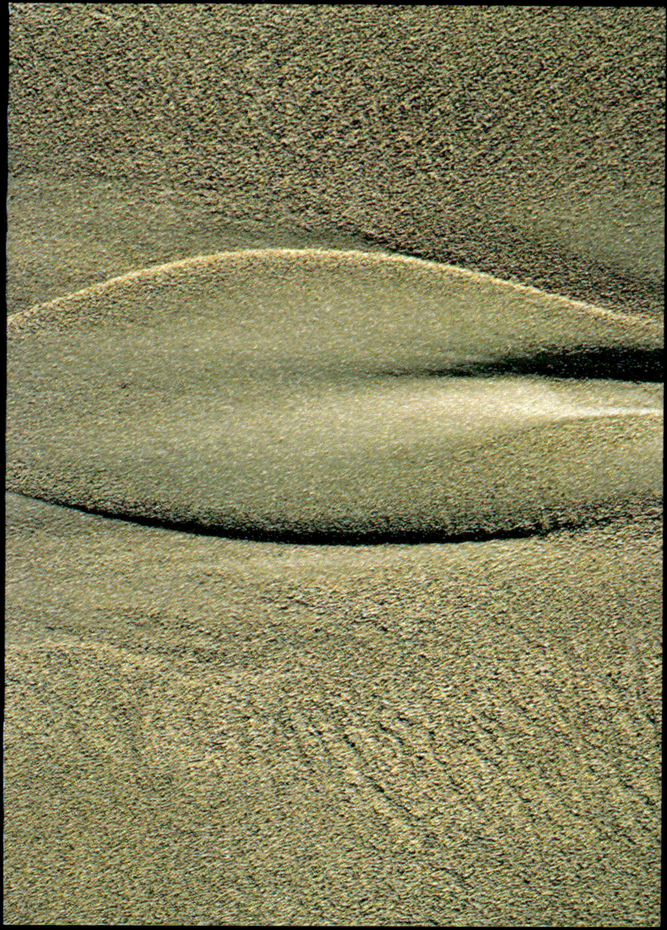
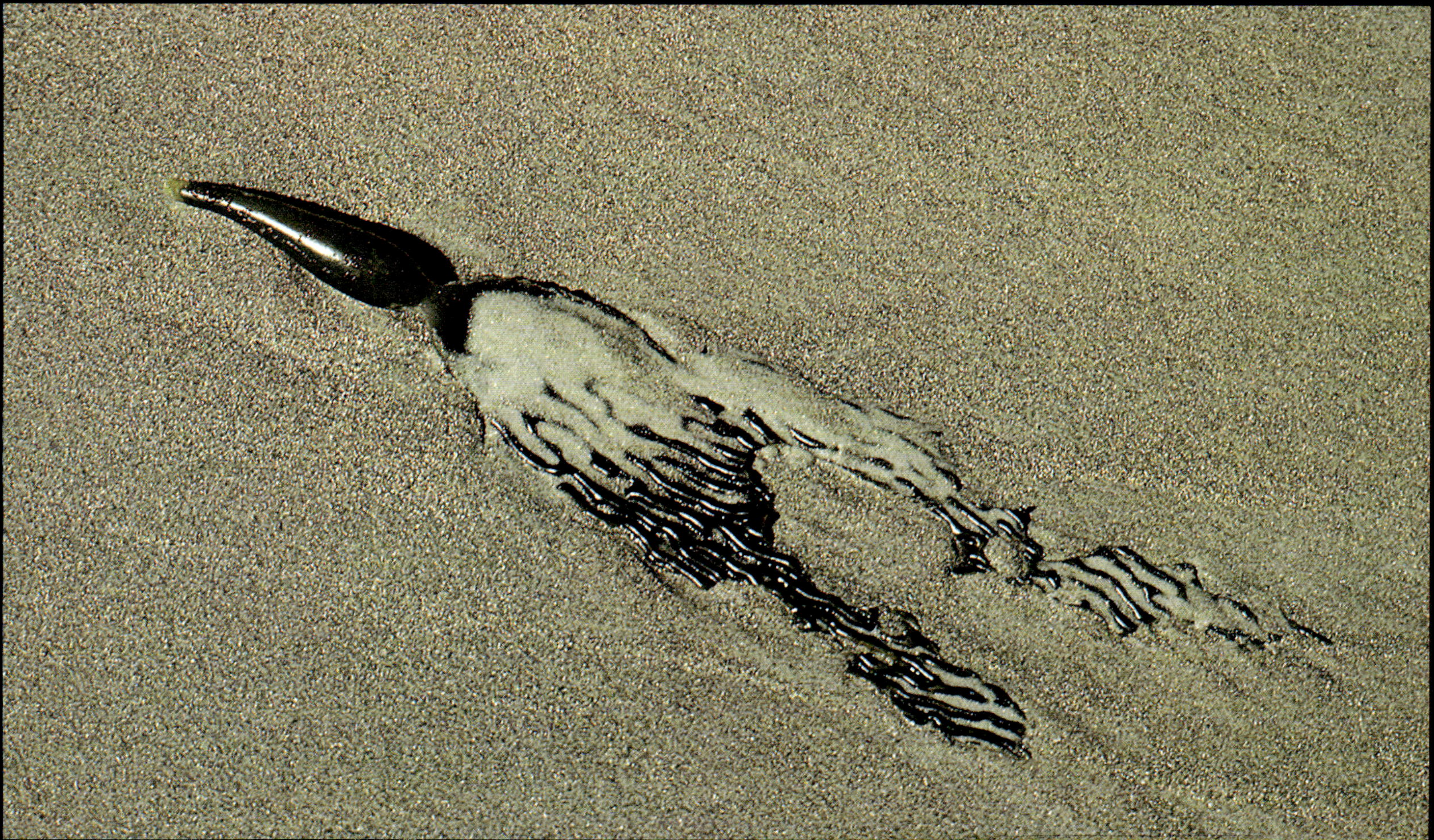

Late afternoon. Mount Karamu is hidden again behind dark cottonwool clouds. To the east the sun highlights the white dolerite peaks of the Amy Range. The sea is calm, its surface barely flexing as swell after swell gently break on shore. It is a deceptive calm as the debris-strewn beach testifies. The tidemark is a morass of rotting kelp (which the tiny sea lice are consuming with a passion), dead shells and driftwood.

Early morning. The rain has retreated to a drizzle. Mount Karamu's peak is still buried under ominous cloud. Out to sea it looks even bleaker. There is no sign of the sun to the east, but inland, patches of blue pockmark the grey-black mass. Late morning, Karamu is still shy.

Two o'clock. We've made the summit – or at least we think it's the summit – we can't see more than 15 metres in any direction. It is an eerie silent world in between wind blasts. The cloud is a thick ooze. There is absolutely no clue to the sun's whereabouts in this strange half light. All plant life is severely stunted and wind-pruned, the natural legacy of the cold westerlies that consistently batter this coastline.

Somewhere out there lies the South West Cape – a shimmer of rock taunting the wild oceans. It was our ultimate target. Now it must remain a dream. And so, in a strange way, this is an end of a journey and a beginning. Tomorrow our path lies constantly east until we reach Cockle Bay – 100 kilometres away.

At Cox Bight, a curious red tide of spongy algae (BELOW AND NEAR RIGHT) leaves a filigree of rotting material on the beaches. It proves no deterrent to a spiny, fur-covered echidna (OPPOSITE) which purposely continues its fossicking nearby.

A sparkling sunset tinges the grassy shoreline of Louisa Bay and takes the edge off the bleakness that characterises so much of this dramatic coastline. The view west from Osmiridium Beach (BELOW RIGHT) provides a rugged panorama.

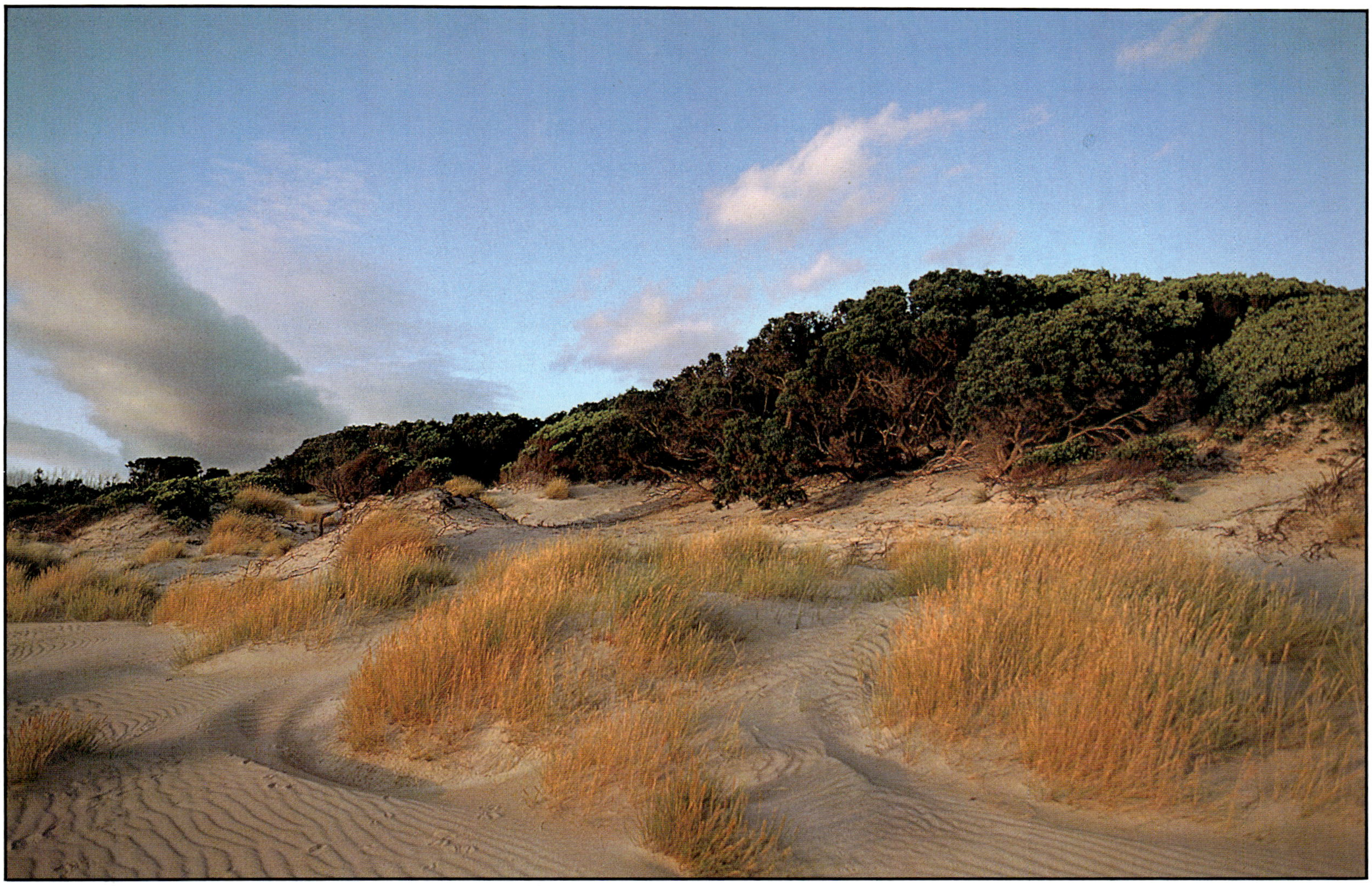

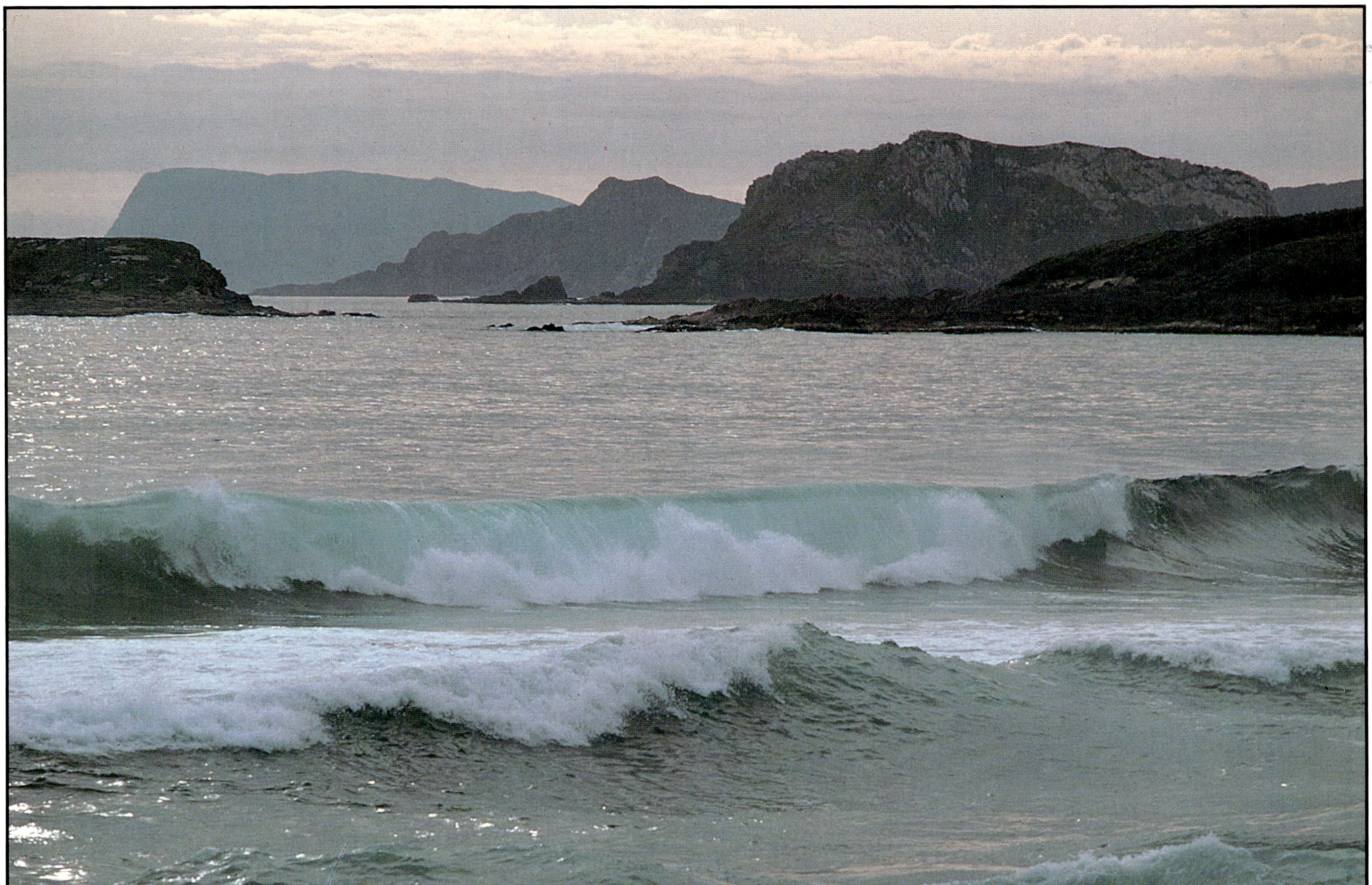

The return journey is a smorgasbord of wilderness memories... sooty oyster-catchers keening on high, grass parrots taking short hops to fresh cover, cockatoos noisily pirouetting overhead, the magnificent wingspread of fish eagles soaring high above, a strange tide of red spongy material at Point Eric, the challenge of the Ironbounds.

And there is the first sunset in two weeks at Louisa Beach. The spiky grass on the sand dunes, a bleached corn colour contrasted against the dim grey of the bulk of the Ironbounds, the patterns etched on the sand, the language of the birds and shells written in the sand...

From the top of the challenging Ironbound Range, the offshore islands are levitated, their bulk floating in a haze creating a seamless horizon. The long downhill stretch through rainforest is wet, muddy and exhausting, but acknowledgement that the steepest climb is over soon revives spirits and the sea beckons for a cold swim.

The South Coast Track gradually feels less and less like a wilderness as we approach Cockle Bay. Daily, there are more people on the track, but it serves as a gradual transition from the untamed territory behind us. As the journey slowly comes to an end, the South West Cape still beckons. Perhaps our journey into wilderness is just beginning.

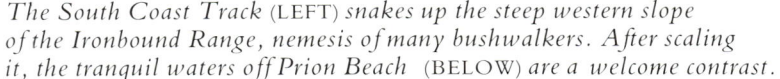

The South Coast Track (LEFT) *snakes up the steep western slope of the Ironbound Range, nemesis of many bushwalkers. After scaling it, the tranquil waters off Prion Beach* (BELOW) *are a welcome contrast.*

Index

The crystal clear atmospheric conditions of the Simpson Desert provide a rare opportunity to observe our universe –
perhaps the truest wilderness of all.